Risk and Resilience in the Era of Climate Change

Vinod Thomas

Risk and Resilience in the Era of Climate Change

palgrave
macmillan

Vinod Thomas
Lee Kuan Yew School of Public Policy
National University of Singapore
Singapore, Singapore

ISBN 978-981-19-8620-8   ISBN 978-981-19-8621-5 (eBook)
https://doi.org/10.1007/978-981-19-8621-5

© The Editor(s) (if applicable) and The Author(s), under exclusive license to Springer Nature Singapore Pte Ltd. 2023
This work is subject to copyright. All rights are solely and exclusively licensed by the Publisher, whether the whole or part of the material is concerned, specifically the rights of translation, reprinting, reuse of illustrations, recitation, broadcasting, reproduction on microfilms or in any other physical way, and transmission or information storage and retrieval, electronic adaptation, computer software, or by similar or dissimilar methodology now known or hereafter developed.
The use of general descriptive names, registered names, trademarks, service marks, etc. in this publication does not imply, even in the absence of a specific statement, that such names are exempt from the relevant protective laws and regulations and therefore free for general use.
The publisher, the authors, and the editors are safe to assume that the advice and information in this book are believed to be true and accurate at the date of publication. Neither the publisher nor the authors or the editors give a warranty, expressed or implied, with respect to the material contained herein or for any errors or omissions that may have been made. The publisher remains neutral with regard to jurisdictional claims in published maps and institutional affiliations.

Cover illustration: acilo/Gettyimages

This Palgrave Macmillan imprint is published by the registered company Springer Nature Singapore Pte Ltd.
The registered company address is: 152 Beach Road, #21-01/04 Gateway East, Singapore 189721, Singapore

To my sister, Malini

Foreword

This authoritative and clearly written book by a renowned economist focuses on the key concepts of risk and resilience in the context of sustainable development under the imminent danger of climate change. Prof. Thomas' rigorous and skillful analysis and policy applications highlight the need for holistic solutions to climate change and other sustainable development issues, as envisaged in the 17 Sustainable Development Goals (SDG).

The book should inspire decisionmakers to formulate new policies and strategies for resilience and sustainability—and avoid the pitfalls. It will help point the way towards a balanced inclusive green growth (BIGG) path and an eco-civilization for the twenty-first century.

The principal challenge facing humanity is sustainable development, which is at great risk as highlighted by the climate crisis. Here, we are all stakeholders because these risks and building resilience to them affect us all. For over 6 decades, mainstream development focused on material-based economic growth, to overcome problems of poverty, hunger, sickness, and inequality. These issues are still severe in most of the poor countries, and even among poorer communities in the rich countries.

More recently, the concept of sustainable development has become more recognized as the framework needed to address multi-faceted and interlinked global issues. An essential element of this approach, discussed in this book, is to seek solutions, including resilience to growing risks, that harmonize and balance the economic, environmental, and social dimensions of the sustainable development triangle. At the global level, the United Nations (UN) has sought to operationalize these ideas through the 2030 Agenda and the SDGs, universally accepted by all countries in 2015.

One vital SDG is climate action, which recognizes that climate change is the ultimate threat multiplier—worsening all the other sustainable development problems. In this context, leading international scientists in the UN Intergovernmental Panel on Climate Change (IPCC) have clearly confirmed that continued human activities which emit greenhouse gases (GHG) lead to catastrophic climate change, with the greatest impacts falling on the poor. This is the major driver underlying "climate justice", since the poor suffer the most, although they have the least responsibility for GHG emissions that heighten the climate risk. The two principles: "polluter pays" and "victim is compensated", argue that the rich will need to transfer resources to the poor to offset climate damages.

The most effective way of tackling the climate conundrum is to integrate climate mitigation and adaptation polices within sustainable development strategies. Maintaining this balance in the sustainable development triangle (economy, society, and environment) dispels the "persistent false dichotomy" between economic prosperity and environmental protection—as argued by Prof. Thomas. The alleviation of poverty among billions requires continued economic progress in these areas, but reducing inequality and poverty must go hand in hand.

The concept of ecological footprint highlights the huge risk associated with humankind overusing ecological resources equivalent to 1.7 Earths. By 2030, it will need the equivalent of two planets to sustain the current way of life. Meanwhile, consumption is highly unequal with the richest 20% of the world's population consuming more than 85% of planetary resources, 60–70 times more than the poorest 20%. Climate risk is rooted in the fact that just one percent of the rich emit 175 times more GHG per person than the poorest 10%.

In this book, Prof. Thomas shows his deep commitment to sustainable development, especially its attention to vulnerable people and groups. As a highly experienced global practitioner in development policy, who has served at the most senior levels of both the World Bank and the Asian Development Bank (ADB), his advice is eminently practical. As a distinguished academic and veteran university professor, with an enviable record of publications and presentations at high level meetings, he presents his arguments lucidly and convincingly.

The book is valuable and highly relevant for global decision-makers, practitioners, researchers, students, and the public. Its attractive features include: a lucid explanation of the key principles linking risk and resilience, and its application to the "intractability of climate change"; clear examples of practical policy applications relevant to a range of circumstances, countries, sectors, and ecosystems, and an extensive and up-to-date bibliography to aid further research.

Colombo, Sri Lanka

Mohan Munasinghe
Chairman, Munasinghe Institute
for Development (MIND)
and MIND Group 2021, Blue
Planet prize Laureate, Shared
the 2007 Nobel Prize for Peace
(as Vice Chair IPCC-AR4),
Chairman, Presidential Expert
Committee on Sustainable Sri
Lanka 2030 Vision and Strategic
Path

Preface

Environmental, social, and economic risks confront countries, the global economy, and all segments of society. Closely linked to risk is resilience, which is getting increased attention from policymakers. With risks rising globally, there is a growing realization that resilience building must go beyond simply strengthening how things are. Building resilience in these uncertain times means re-starting from an improved position. Acting on resilience with anticipation would represent a sea change in development practice.

Climate change is the gravest development risk facing humanity. Its dangerous trajectory coincides with the COVID-19 pandemic, the biggest health crisis in a century, and Russia's war in Ukraine that began in early 2022 and triggered severe food shortages in parts of the world. Decades of environmental destruction and high carbon economic growth were spearheaded by industrialized countries and followed by emerging economies. Backed, implicitly or explicitly, by mainstream economic policy and advice, they are the roots of today's rapidly unfolding climate crisis. The climate problem exacerbates other daunting challenges of shrinking water resources, financial stress, biodiversity extinction and involuntary migration.

Risk and Resilience in the Era of Climate Change brings together frameworks of analysis and the latest findings on the dangers of and responses to runaway climate change and other rising threats. The risks to lives and livelihoods, the accelerating threat to biodiversity and the very health of the planet will demand far-reaching changes in behaviour and sizable investments in resiliency—and a global consensus of zero tolerance for climate inaction. We are very far from both. Countries are grappling with the hazards of climate disasters, but because the crisis is global, the threat can only be effectively mitigated by including the international response. Beyond climate disasters, economic agents are also worried about stranded assets, supply chain bottlenecks, and financial stress.

The high and rising stakes do not argue against all risk taking, for example, in finding innovative solutions to climate change itself. But they revise the felt need for resilience building, especially in countries prone to hazards of nature, to have the capability and capacity to counter the next disaster. Resilience building tries to add flexibility and dynamism to tackle new dangers. A slew of economic tools and regulations is available for shoring up resilience, for example, for targeting carbon emissions. But all these have yet to have anything close to the impact needed to stave off the most deleterious effects of climate change. Trends in GHG emissions—notwithstanding some decoupling of Gross Domestic Product (GDP) and emissions in several rich countries and a brief respite from the COVID-19 pandemic—have continued to move in the wrong direction. The situation calls for an overhaul of global governance that will enable an enforcement of commitments needed to achieve global climate goals. Averting catastrophes through such efforts would not mean that the original predictions were false.

Understanding the risks of climate inaction and the need for resilience building tools that are most fit-for-purpose can help build the political support for action that is needed to avert the worst effects of global warming. Risk and resilience frameworks can help formulate climate policies that are suitable for country situations. Beyond the specifics of policy measures is the crucial change required in the mental frame in policymaking from seeing climate activity as an impediment to economic growth to an essential aid to sustained growth and wellbeing. Indeed, a better approach is needed for how economic growth is conceived and measured in the era of climate change. The UN's SDGs target the quality of growth, and this book underscores the value of so doing. Conference of the Parties (COP) 27 and its follow up must record greater commitments of countries on climate action.

This work adds to a growing literature not only on risk and resilience but also on how solutions to the climate conundrum might be found. The work brings together the essentials of analyses and experiences that readers can draw on to arrive at conclusions. But it goes further in offering key differences in observations and implications that could help move the needle on getting better results. That leaves room for innovations, ingenuity, and surprises along the way that aid in the search for solutions. The book grew out of teaching graduate classes on climate and environmental policy, and development management at the National University of Singapore and at the Asian Institute of Management. It draws on three decades of operational work, policy discussions, and research at the World Bank and ADB. That involved working with countries and the international financial community, including a decade spent evaluating the development effectiveness of investment projects and programs in all regions, particularly regarding their environmental sustainability and climate impacts.

Singapore, Singapore   Vinod Thomas

Acknowledgments

I would like to acknowledge the substantial contributions of several colleagues in writing this book. I thank Lee Kuan Yew School of Public Policy research assistants Chitranjali Tiwari and Kendra Wong for their excellent research and presentational improvements during the project, and Saba Abdulaziz S. Altwayan, Marion Hill, Monica Khoo, Shameen Idiculla, Megan Leong, Aman Thomas, Leila Thomas, Tsai Yi-Chen, Ranjana Sengupta and Shih Hui Voon towards its end. Detailed and useful feedback from Laveesh Bhandari, Deepak Bhattasali, Keith Cabaluna Detros, Siong Guan Lim, Igor Lincov, Robert Picciotto, Euston Quah, Jose Sokol, Anil Sood, Daniel Rajasingam Subramaniam, Benjamin D. Trump, and Shahid Yusuf is deeply appreciated. I would like to note my close association with the work of Ramon E. Lopez. This work benefited from collaboration with Eduardo Araral Jr, Amar Bhattacharya, Benjamin W. Cashore, Lesley Y. Cordero, Gautam Kaji, Jikyeong Kang, Vikram Khanna, Muthukumara S. Mani, Vikram Nehru, Danny Quah, Marqueza Cathalina L. Reyes, and Eng Dih Teo. Support from Institute of Water Policy, and Institute for Environment and Sustainability, and collaboration with Asian Institute of Management are gratefully acknowledged.

Praise for *Risk and Resilience in the Era of Climate Change*

"As we stare down at the climate emergency, this publication provides crucial insights into risk and resilience in troubled times. The work enriches our understanding of smart policy instruments which serve a dual aim, to mitigate climate change and enhance resilience. In so doing, it makes a key contribution to viewing economic growth through a different lens, one which accounts for the quality of growth, resilience, social inclusion, and environmental sustainability—where climate action makes sustained growth possible."
—Inger Andersen, *Under-Secretary-General of the United Nations, and Executive Director of the UN Environment Programme*

"Professor Vinod Thomas has eloquently and forcefully highlighted how important it is to improve climate resilience in the energy sector. He provides invaluable insights on the impacts climate change is already having on energy systems and about the key actions that are needed in response, covering policy, communication, innovation, investment, and individual behaviour. By doing so, he delivers a strong message that building energy sector resilience against climate change requires a multi-faceted approach across all parts of society."
—Fatih Birol, *Executive Director of the International Energy Agency*

"With sweeping coverage and compelling analysis, Vinod Thomas explains the rising risk from climate change and other threats facing society. But Professor Thomas does not stop there. Rather he highlights the path forward including the need for investments in resilience, better disaster management, recast economic thinking, improved global governance, business transformation, and most urgently, a shift in mindsets that will reshape education, public values, and ultimately behavior. Critical reading for policymakers, corporate leaders, environmental advocates, energy experts, as well as students and scholars – and anyone else interested in a sustainable future."
—Daniel C. Esty, *Hillhouse Professor, Yale University and Editor*, A Better Planet: 40 Big Ideas for a Sustainable Future

"With the sharp rise in climate risks, various discussions of policy and technical solutions are at center stage. However, for these deliberations to have meaningful traction, they need to be supported by a sea change in people's mindsets and policymakers' priorities for action. This book makes a significant contribution in calling for transformational change in values, education, and behavior that will underpin the sustainability of development. A must-read for policymakers and academics alike."
—Jikyeong Kang, *President, Dean and MVP Professor of Marketing, Asian Institute of Management*

"The COVID-19 pandemic is a stark reminder of humanity's vulnerability to shocks from nature. The costs to humanity depend on how well we understand the nature of the shock, take actions in advance for mitigation and adaptation, and strengthen society's resilience with preparations. Among all known possible shocks from nature, climate change driven by global warming is likely to be the most threatening. This informative and authoritative book by Professor Vinod Thomas provides convincing arguments and practical advice for countries in the north and the south to build the needed resilience before it is too late. This book is timely, and a must read for anyone concerned with humanity's future."
—Justin Yifu Lin, *Dean, Institute of New Structural Economics, Peking University, Former Chief Economist and Senior Vice President, World Bank*

"Vinod Thomas's urgent call to action on climate change is both well-informed and well-intentioned, providing clear and compelling arguments for reforming global governance and sharply shifting development priorities from economic growth to climate change. It is indispensable reading for policymakers, leaders of development organizations, and all those committed to building a sustainable future."
—Albert Park, *Chief Economist, Asian Development Bank*

"This timely work on risk and resilience in the world of climate change is an important one. In addressing climate change, we would do well to construct better models of resilience, understanding types and degrees of risk estimates, and not be bogged down with just economic efficiency arguments, and neither should it be based on only socio-economic considerations. Professor Vinod Thomas, with his relevant and wide background in economics and public policy, both in the academia and in the real policy front is in a very good position to combine these important elements to push further our understanding of climate change issues. I fully endorse this book."
—Euston Quah, *Albert Winsemius Chair Professor and Head of Economics, Nanyang Technological University, and Editor, Singapore Economic Review*

"As crises hammer our world with increased intensity and frequency, it is urgent that leaders and policymakers read this book. Professor Vinod Thomas brilliantly demonstrates, with abundant data and evidence, that addressing the climate change crisis urgently and holistically makes good economic sense and that this will be best done by strengthening resilience at the local, national, regional, and global levels. As the house burns, this book can convince even the most skeptical that there will be no future if the risk posed by climate change is not better understood and addressed, and if resilience is not placed at the very center of development strategies."
—Veronique Salze-Lozac'h, *Chief Evaluator, European Bank for Reconstruction and Development*

"Everyone should read this clear-eyed and expertly-researched analysis of the role climate change is playing and will play in human lives. Dr. Thomas walks us through the science, the public policy, the need to both mitigate and adapt, the strategies and best practices, and how we can transform the trajectory. Amongst other important concluding recommendations, Dr. Thomas issues a clarion call for moving away from our overuse of GDP to design public policy objectives that prioritize growth over societal impacts. To paraphrase Dr. Thomas – it is time to start measuring quality over quantity and make our personal, professional, and political decisions based on improving the quality of life for all."

—Tensie Whelan, *Professor and Founding Director NYU Stern Center for Sustainable Business*

"This book shows with great clarity and by drawing upon recent publications, that there is an urgent need to recognize the increasing risk posed by climate change, to alter mindsets, and to fundamentally rethink economic strategies. If growth is to remain a priority, it must be rendered sustainable through the avoidance of diseconomies and by measures that build resilience in the face of shocks to come. This compact and incisive volume is a welcome addition to the literature on a vital topic. It could nudge the undecided to recognize the enormity of the challenge that is upon us."

—Shahid Yusuf, *Chief Economist, Growth Dialogue and Non-Resident Fellow, Center for Global Development*

Contents

1 Opening Summary   1
    Bibliography   7

Part I   Risk and Resilience

2 Troubled Times   11
    New and Extreme Danger   12
    The Nature of Risk   16
    The Significance of Resilience   24
    Climate Risk and Resilience   26
    Conclusions   29
    Bibliography   29

3 Understanding Risk   35
    The Global Risk Landscapes   36
    Risk and Uncertainty   39
    Measuring Risk   40
    Sources of Disaster Risks   44
    Conclusions   48
    Bibliography   49

4 Resilience That Shapes Risk   53
    Nature and Phases of Resilience   54
    Interaction Between Risk and Resilience   57
    Mapping Components of Risk and Resilience   61

Country, Regional, and Global Priorities	64
Conclusions	68
Bibliography	68
5 New Highs in Risk and Resilience	**73**
Rising Risks	74
Anchoring Resilience	78
Conclusions	87
Bibliography	88

Part II The Climate Catastrophe

6 Intractability of Climate Change	**95**
Unheeded Warnings	96
Problems Eluding Solutions	98
A Super Wicked Problem	103
Lines of Causation	107
Adding Fuel to the Fire	109
Failure in Messaging	113
Conclusions	115
Bibliography	116
7 A Persistently False Dichotomy	**125**
Centrality of Externalities	126
Discounting the Future	131
Econometrics and Climate Disasters	133
Economics at the Climate Table	136
Conclusions	137
Bibliography	138
8 Integrating Resilience in Policymaking	**143**
Crisis and Disaster Management	144
Disaster Prevention and Mitigation	151
Mitigation and Adaptation Efforts	152
Alternative Energy Sources	158
Costs and Benefits	159
Green Financing	161
Sources of Climate Funds	163
Conclusions	165
Bibliography	166

9 Transformative Change	**173**
The Big Picture	174
Triage Financing and Pricing	177
Energy Transition	179
Global Economic Policymaking	181
Societal and Individual Behaviour	183
Conclusions	189
Bibliography	190
Index	**197**

ABBREVIATIONS

ADB	Asian Development Bank
AIIB	Asian Infrastructure Investment Bank
ASEAN	The Association of Southeast Asian Nations
BRI	Belt and Road Initiative
C	Celsius
CCL4	Carbon Tetrachloride
CFCs	Chlorofluorocarbons
CH4	Methane
CO2	Carbon Dioxide
COP	Conference of the Parties
CSA	Climate-Smart Agriculture
CVAR	Cointegrated Vector Autoregression
DU	Dobson Unit
EU	European Union
FAO	Food and Agriculture Organization
FiT	Feed-in Tariff
GDP	Gross Domestic Product
GHGs	Greenhouse Gases
GPOD	Global Probability of Disaster
GSDRC	Governance and Social Development Resource Centre
HFC	Hydrofluorocarbons
IAMs	Integrated Assessment Models
ICU	Intensive Care Unit
IEA	International Energy Agency
IMF	International Monetary Fund
IPCC	Intergovernmental Panel on Climate Change

IPPs	Independent Power Producers
IUCN	International Union for Conservation of Nature
JBIC	Japan Bank for International Cooperation
MDB	Multinational Development Bank
MERS	Middle East Respiratory Syndrome
N2O	Nitrous Oxide
NASA	National Aeronautics and Space Administration
NF3	Nitrogen Trifluoride
NOAA	National Oceanic and Atmospheric Administration
OECD	Organization for Economic Co-operation and Development
OPEC	Organization of the Petroleum Exporting Countries
PFC	Perfluorocarbon
PM	Particulate Matter
RAFT	Resilience Adaptation Feasibility Tool
SARS	Severe Acute Respiratory Syndrome
SDGs	Sustainable Development Goals
SEEA	System of Environmental-Economic Accounting
SF6	Sulphur Hexafluoride
UK	United Kingdom
UNDRR	The United Nations Office for Disaster Risk Reduction
UNEP	United Nations Environment Program
UNFCC	United Nations Framework Convention on Climate Change
US	United States
WDR	World Development Report
WEF	World Economic Forum
WFP	World Food Programme
WHO	World Health Organization
WMO	World Meteorological Organization

List of Figures

Fig. 2.1	Hazards of nature on the rise (*Source* Buchholz 2020; Munich RE 2022)	14
Fig. 2.2	Carbon emissions and temperatures (*Source* Andrew and Peters 2021; Global Carbon Project 2021; NASA 2022; NOAA 2022)	20
Fig. 2.3	Comparing risks of climate change and the pandemic (*Source* Based on Gourinchas 2020; Verkooijen 2020)	23
Fig. 2.4	The prevention–response axis of disasters (*Source* Taken from World Bank 2013)	24
Fig. 3.1	Perceptions of global risk (*Source* Based on World Economic Forum 2021)	36
Fig. 3.2	Profiling risk (*Note* The arrows show the impact and the probability going from low to medium to high, and the cells represent low, medium or high risk. *Source* Based on well-known depictions)	41
Fig. 3.3	Contributors to disaster risk (*Source* Kahan 2010; UNISDR 2017; Author's adaptation)	44
Fig. 4.1	Components of building resilience (*Source* Author's depiction drawing on a vast literature)	56
Fig. 4.2	Cases of improving resilience (*Source* Author's depiction)	58
Fig. 4.3	Risk, resilience, and system functionality (*Source* Based on Linkov et al. 2013 and MDBs' project completion reports)	60
Fig. 4.4	Interaction of risk and resilience (*Source* Based on Linkov and Trump 2019 and the authors' presentations)	62

Fig. 5.1	Resilience metrics (*Source* Based on Linkov et al. 2013)	79
Fig. 6.1	Number of endangered species is rising (*Source* Buchholz 2021)	99
Fig. 6.2	The climate connection (*Source* Author's depiction; The Royal Society 2022)	100
Fig. 6.3	Stratospheric Ozone Concentration (mean) (*Source* Our World in Data 2022)	105
Fig. 8.1	Phases of a disaster (*Source* Author's depiction based on the literature)	145
Fig. 8.2	Mitigation and adaptation (*Source* Taken from Parry [2009])	153
Fig. 8.3	Shifting the risk-reward profile or a renewable project (*Source* Based on Glemarec et al. [2012])	162
Fig. 9.1	Aggravation of the climate crisis (*Source* Author's depiction)	183

List of Tables

Table 5.1 Preventive action in Odisha, India (indicative orders
          of magnitude)   82
Table 5.2 Superstorms in the Philippines (rough and indicative
          estimates)   83
Table 6.1 Emissions by Country   110
Table 7.1 Prices, interest rates and externalities   133

# List of Boxes

Box 2.1	Greenhouse Gases and Carbon Emissions	18
Box 2.2	Carbon Neutrality and the Paris Agreement	28
Box 3.1	Risk and Resilience in Cybersecurity	37
Box 4.1	Social Capital and Response to the Nepal Earthquake 2015	55
Box 4.2	Resilience Building in Towns	59
Box 4.3	Global Food Insecurity and the Russia-Ukraine War	65
Box 4.4	Public Health Systems and Resilience	67
Box 5.1	Kenya Climate-Smart Agriculture Project (CSA)	81
Box 5.2	China's Belt and Road	86
Box 6.1	Delhi, Air Pollution and Transportation	106
Box 7.1	Mangroves or Shrimps?	126
Box 7.2	Climate Change and Prices	132
Box 8.1	Vietnam's Resilient and Decarbonising Pathways	145
Box 8.2	Mozambique Sustainable Irrigation Development Project	150
Box 8.3	Singapore and Carbon Tax	156
Box 9.1	The 2022 US Climate, Health, and Tax Bill	178

CHAPTER 1

Opening Summary

Knowledge is a free good. The biggest cost in its transmission is not in the production or distribution of knowledge, but in its assimilation. Kenneth Arrow

Aggregate risks in the early 2020s are extremely high. That makes it a crucial time to assess risk and resilience and to do so from three perspectives—the environment, society, and the economy. The confluence and cumulative effects of the COVID-19 pandemic, Russia's invasion of Ukraine in February 2022, climate catastrophes, and the economic shocks and food shortages that they unleashed, are sharp reminders of the vulnerabilities to risks and the urgent need to build and scale up resilience.

The essential question is whether these events are short-term or whether they are an indication of worse things to come, especially if they are the tip of the iceberg. How risks are viewed makes a difference to mindsets and policies on resilience, i.e., the ability to bounce back from setbacks. In the policy context, risk is the eventuality of a worse outcome compared with what is expected. If the risk trajectory is high and rising, resilience building is not only about recovering but also about investing in readiness. It calls for serious and meaningful change.

Many of the principles and lessons of risk and resilience apply to disasters in general, including natural hazards, pandemics, cyber threats, food shortages, and financial crises. The special focus of this book is resilience against climatic extremes. The spotlight on climate change leads to a

© The Author(s), under exclusive license to Springer Nature Singapore Pte Ltd. 2023
V. Thomas, *Risk and Resilience in the Era of Climate Change*,
https://doi.org/10.1007/978-981-19-8621-5_1

scenario of increasing certainty that the danger is rapidly rising. While people now flag climate change as a top concern, they do not put climate investment above many other expenditure categories, which is the test of true priorities. The top priorities on the minds of voters in the United States (US) and elsewhere do not include the environment and climate change. Resilience building in these respects will require a radical shift in this mindset.

Seven cross-cutting messages emerge from this work:

- Attributing damage from weather extremes to its root cause viz carbon emissions related to historic and current human activity, is vital to shaping public opinion and policies. It does not help when news anchors report extreme weather events as the work of Mother Nature. Climate change is an extreme example of a gulf between scientific knowledge about the risk and economic policies needed to deal with it. The tobacco-cancer experience suggests that establishing the link between cause and effect is key to accountability. In the case of global warming,[1] demonstrating causation begins with the correlation and scientific linking of temperatures and carbon emissions. To motivate mitigation, disseminating the new "attribution" reports causally connecting fossil fuel-based energy and extreme weather events will help. Increasingly, the public labels events as climate disasters, but this is not enough: they need to be understood as human-caused and avertable episodes (Chapters 2–4).
- Policy decisions are predicated on communicating the linkages effectively. For climate change, descriptions of the disasters it causes are plentiful, and so are stories of bravery in withstanding the tragedy—e.g., in daily weather reports, and descriptive television coverage of every detail of an event. But there is a lack of recognition of the link between disasters and emissions, especially in the middle of a storm. This represents a great breakdown in communication, in connecting cause and effect. Where messaging has been strong, the scenarios have focused on the future, depicting the impacts as distant in time and space, rather than here and now. Clear communication is fundamental because the links are indirect (fossil fuels-emissions-global warming-disasters) rather than direct as for, say, COVID-19 (infection-hospitalisation) (Chapters 3, 4 and 6).

- Climate change calls for recognising a shift in the nature of the risk from a low-probability but high-impact situation to a high-probability and high-impact one. This type of change in risk affects all phases of the disaster management cycle—preparedness, relief and response, and reconstruction and recovery. But the more predictable the increase in extreme disasters, the greater is the premium on readiness and preparedness. Political will is crucial to taking a preventive stance because its rewards are not necessarily visible immediately. In 2022, the US Senate finally passed a landmark bill, the Inflation Reduction Act, dedicating significant resources for climate and energy investment. The significance here is that leading by example is necessary to galvanise a global movement and produce effective climate action (Chapters 3–5).
- Mainstream economics has, ironically, been unhelpful in the policy-making that is needed to combat climate change. Economic analysis has been instrumental in promoting investments in health and education, but it has failed to anticipate the rapidly escalating climate risk. Its preoccupation with short-term GDP has been at the expense of a focus on negative externalities or negative spillovers from human activities—like carbon emissions and deforestation—which diminish the net domestic product. The false belief that environmental protection is inimical to sustained growth needs to be discredited, so that decision makers can move away from unduly discounting the benefits of climate investment that accrue in the future. All countries should adopt carbon pricing as a minimum step to discourage carbon-intensive growth (Chapters 7 and 9).
- With climate change, risk reduction goes beyond coping with people's exposure and vulnerability to a shock to building back better and "bouncing forward" (Manyena et al. 2011). Reducing risks needs to include dealing with the direction and intensity of the shock itself. Resilience building, in this case, entails more than securing capabilities to live with the inevitable; it also needs to incorporate efforts to reduce the intensity of the hazard by decarbonising economies and reducing the rise in temperatures. Climate resilience calls not only for adaptation that enables a better adjustment to disasters, for example, building sea defences, but also mitigation to reduce disaster risks by cutting greenhouse gases (GHGs) through a phase-out of fossil fuels, and avoiding the kind of expansion of coal seen in 2022–23 with new plant approvals and exports (Chapters 4 and 8).

- Heavy lifting will be needed in innovating approaches to resilience as all countries or localities face shortages in staff and financial resources—a lesson learnt during the COVID-19 pandemic. Great value is to be had in mobilising resources across boundaries, stepping up efforts, and innovating beyond norms. It is especially important to do so ahead of calamities, as precious time will be lost if fund-raising waits for them to strike. Climate change also warns of downward spirals, for example, extreme weather hurting energy supplies, rising energy prices, and the pressure to expand fossil fuels—all of which aggravate global warming. On the other hand, the demand for renewable energy sources like solar energy could rise, and far more resources will need to be deployed to bring them online. Economic growth helps financing; but for it not to be counter-productive, growth should be net zero carbon (Chapters 5, 8 and 9).
- The priority for investing funds in climate resilience needs to be urgently and vastly raised because the price of delayed action is rising sharply. The urgency for pre-disaster prevention as well as efforts to re-build smarter become elements of a paradigm shift and transformational change. Spending on climate imperatives needs to accelerate, but not at the expense of the quality and efficiency of that spending, which in turn are needed to attract financing. All investment projects should require climate proofing, for example, ensuring resistance to extreme weather. Countries would want to carry out climate stress tests under scenarios of increasing risks, much as the central banks regularly do stress tests to assess the health of financial systems (Chapters 1 and 9).

This book is about the constituents of risk and resilience in their multiple facets but is driven by fundamental qualifications that emerge from its primary focus on human-made climate change (Kolbert 2022). It is informed by a distillation of the lessons of experience from independent evaluations of thousands of development projects in all regions of the world over the past decades. The lessons remind policymakers and practitioners of the value of deriving findings even as projects and programmes are ongoing, and not just waiting until all the results are in. They also suggest that past lessons may not be a sufficient guide for the future in the presence of rapidly changing circumstances.

Behind the environmental, social, and economic components of the problem covered in the book are deep-seated political economy considerations. A "super wicked" problem featuring contradictory and intractable features that are hard to solve, climate change has antecedents in history—including the presumption of a fast rising human population that it could mortgage the earth. It presents a conundrum of the worst kind involving a global public good. People increasingly label climate change as the biggest developmental danger, but most do not rank it as the top actionable item in the political economy of reforms. The calculus of political leaders in turn displays priorities for shorter time horizons that miss the climate imperative. While the focus of this book is policies and investments that are needed to build better resilience against rising risks, these steps link back to people's and policymakers' understanding, attitude, and behaviour regarding risk and resilience.

While risk and resilience are discussed in relation to climate change, a game-over threat unless urgently confronted, building resilience is needed for many other risks with which the world is already grappling—pandemics, food shortages, economic and financial shocks, cyber security, and geopolitical tensions. These risks, also wicked problems in varying degrees, are also competing for stretched resources. Often, they have common roots and crisscross each other; their solutions often feature synergies as well. Indeed, many of the underpinnings, discussed primarily for the climate case, also apply to other wicked problems.

For the climate crisis and other extreme problems, such as food scarcity, building resilience is no longer just about recovering but re-building better in anticipation of higher bars to cross. That vast sums can be quickly mobilised to fix global problems was dazzlingly demonstrated in the trillions of dollars—US$15 trillion in 2020 by the so-called G-10 plus China in one estimate raised to fight COVID-19 (Wilkis and Carvalho 2020). Yet the world is struggling to raise the US$100 billion annually in climate finance for developing countries for mitigation and adaptation (OECD 2022) as promised in COP15 in Copenhagen. Despite widespread acceptance that climate change is a pressing crisis, the lack of financing shows that it is still being seen as a crisis that lies over the horizon and can be fixed by coming generations.

Solar energy capacity is expanding globally, even if from a small base, spurred by technological innovation, including in batteries. Adoption of breakthrough technologies, for example, green hydrogen or carbon capture, and work on geoengineering options call for investments to enable these possibilities to go commercial. Green hydrogen, from electrolysis to split water into oxygen and hydrogen using renewable power, offers hope as new investments are underway in several countries (Mundle 2022). Nuclear reactors can provide the heat and electricity for clean hydrogen, though costly and years in the making for commercialisation. The fuel could potentially be employed in steel, cement, fertilisers, cooking and heating, transportation , and power. Carbon capture, another hope for the future, would clone the photosynthesis in nature to absorb carbon, the technology for which also exists. The challenge is to take these ideas to scale and make them viable, quickly enough to make a difference. It is hoped that this book will provide essential frameworks and stimulate discussions on raising the bar for resilience amid rapidly rising risks.

Risk and Resilience in the Era of Climate Change is divided into two parts. Part I examines risk and resilience in the world of today; Part II considers its application to climate change. The following chapter examines socioeconomic and ecological risks. It stresses the importance of recognising the anthropogenic and systematic nature of climate-related disasters. Chapter 3 discusses ways of understanding and anticipating risk in the context of extreme threats, like climate change. It emphasises that these risks cannot be separated from socioeconomic considerations. Chapter 4 reviews the role of resilience in the face of exposure, vulnerability and intensity of hazards, and argues for seeing resilience as an opportunity for better postcrisis positioning. Chapter 5 builds on Chapter 2 by looking ahead, and making clear that building resilience calls for innovation and dynamism, going beyond increasing the capacity to recover from disasters.

In Part II, Chapter 6 links with Chapters 2 and 5, and sets out the intractability of climate crisis, making it a "super wicked problem" for which timely solutions have been elusive thus far. It emphasises the role of accountability and messaging. Chapter 7 explains why, ironically, climate change has taken a back seat in economic analysis and policy, that are obsessed with short-term GDP growth. Bringing economics and science together can help in better communication and finding solutions. Chapter 8 presents the challenge of dealing with disasters associated with global warming from the perspective of policy. It draws on useful lessons learnt from the hits and misses of local, national and global actions for climate mitigation and adaptation, and for raising climate finance.

Chapter 9 discusses the need for far-reaching shifts in priorities, attitudes, and mindsets to manage risks and resilience better and tackle climate change as a key part of agendas for sustainable development.

Note

1. Global warming is mentioned with reference to rising temperatures, the centrepiece of climate change, which also manifests itself in melting glaciers, heavier rainstorms, longer droughts—and extreme cold weather. Climate refers to changes in the atmosphere over a period of time, while weather relates to short time periods. Climatic refers to climate and its patterns.

Bibliography

Kolbert, Elizabeth. 2022. "A Vast Experiment." *The New Yorker*, November 28.
Manyena, S.B., Geoff O'Brien, Phil o'Keefe, and Jo Rose. 2011. "Editorial. Disaster Resilience: a Bounce Back or Bounce Forward Ability?". *Local Environment* 16 (6): 1–8. May. https://www.researchgate.net/publication/283681542_Editorial_Disaster_resilience_a_bounce_back_or_bounce_forward_ability.
Mundle, Sudipto. 2022. "Here's the "Steam Engine" of the Twenty-First Century." Livemint, October 21. https://www.livemint.com/opinion/online-views/does-a-bright-future-lie-beyond-the-gloomy-economic-horizon-11666286368601.html.
OECD. 2022. "Climate Finance and the USD 100 Billion Goal—OECD." https://www.oecd.org/climate-change/finance-usd-100-billion-goal/.
Wilkes, Tommy, and Ritvik Carvalho. 2020. "$15 Trillion and Counting: Global Stimulus so Far." *Reuters*, May 12, 2020, sec. Business News. https://www.reuters.com/article/uk-health-coronavirus-cenbank-graphic-idUKKBN22N2EP.

PART I

Risk and Resilience

CHAPTER 2

Troubled Times

An economy may be in equilibrium from a short-period point of view and yet contain within itself incompatibilities that are soon going to knock it out of equilibrium. Joan Robinson

Risks to lives and livelihoods are headline news in every corner of the world, captured most dramatically by images of heatwaves, forest fires, and droughts. The top hottest ten years in recorded history have all been since 2010 (Levitan 2022), as climate change worsens the risk profiles of countries and raises an urgent need to step up resilience efforts. Dangers in this century include pandemics, food insecurity, and Russia's war in Ukraine. While the range of country problems is broad, a growing number of them are increasingly connected to climate change. Although it is debatable whether these types of risks have been rising systematically over the span of centuries, the confluence of more intense shocks in the twenty-first century is striking.

Despite mounting scientific evidence of the risks, the causal relation between carbon emissions, global warming, and hazards still appears intangible to many, as well as being distant in time and space. Remarkably, there is an alarming gulf between the scientific knowledge on climate change and the policies for economic growth needed to arrest global warming. Historical experience could not possibly have prepared society for ever deadlier climate events, like Hurricane Ian on the Atlantic coast, the fiercest to strike the state of Florida since 1935, and the epic floods in

© The Author(s), under exclusive license to Springer Nature
Singapore Pte Ltd. 2023
V. Thomas, *Risk and Resilience in the Era of Climate Change*,
https://doi.org/10.1007/978-981-19-8621-5_2

Pakistan, the deadliest in the country's history, both in 2022. A clearer understanding of the heightened risks and policy responses for building resilience is urgently needed.

When beset by pandemics, financial distress, and extreme weather, anxieties on how to cope run high. Various reports consider perceptions and measures of global risks, typically covering economic, social, environmental, and technological shocks (for example, IMF 2022; WMO 2021; WEF 2022, Pulwarty et al. 2022). Broadly speaking, risk means that there is a probability that the outcomes will differ from the expected ones, especially on the negative side of the ledger. In finance, measures of risk inform the uncertainty that an investor is willing to accept to realise a gain from an investment.

New and Extreme Danger

Environmental and climate risks, the primary focus of this book, have perennially been at the top of lists of global risks, even if actions have not matched this concern. Preventing an acceleration in climate disasters is the top development policy priority, as all else depends on it. Rich and poor, technologically advanced and developing nations are all being hit with destructive ferocity. It is already too late to shield countries, as policies for economic growth have failed to integrate the climate reality. Encouraging boutique examples, notwithstanding, big business continues to make money by burning high-carbon fuels. Climate change is an existential danger since it spans the immediate and longer terms with devastating and increasing intensity.

The global economy, at the time of writing, is facing other severe risks as well. Widening geopolitical fractures are increasingly dictating global fortunes. And related to this is the erosion of social capital, with roots in growing income disparities that were worsened by the COVID-19 pandemic. The effects of COVID-19 on public health and economies continue to be a critical global threat. The initial global recovery from the economic impact of the pandemic has been fragile. The inflationary impact of commodities is volatile and unpredictable. The global economy faces an array of tough macroeconomic risks (Anstey 2022).

The continued emission of GHGs has warmed the Earth, causing more extreme rains and storms, the cardinal signature of climate change. Significant warming in recent times began around 1985, and each decade has been warmer than the previous since the 1980s. Since 2010, warming has hit record highs, with the warmest eight years in recorded history being all from 2015. The hottest years were 2016, 2019 and 2020, and July 2021 was the hottest month ever. Unusually extreme cold and snowstorms in winter are also associated with climate change that is making the polar vortex stretch south (Chapters 5 and 6). Studies from National Aeronautics and Space Administration (NASA) show that accelerated warming is significantly attributable to human causes, with the IPCC's Fifth Assessment Report attributing over half of observed global temperature increase to carbon emissions resulting from human activity from 1951 to 2010 (Wiedmann et al. 2022). Differing analytical views, even if a small minority, are pinned inter alia on the dominance and uncertainty of natural climate variability rather than anthropogenic changes (for example, Curry 2023).

Since 1960, the world has seen a tenfold increase in extreme hazards of nature, with the impacts proving to be deadliest in low-lying coastlines, such as those in southeast US, and in regions that are highly exposed to heatwaves, such as the European Union (EU) (Institute for Economics and Peace 2020; IPCC 2022; Iberdrola 2019). Between 1970 and 2000, reports of medium- and large-scale disasters averaged 90–100 a year, but from 2001 to 2020, the reported number of these events increased to 350–500 a year. Annually reported disasters have increased significantly in the last two decades (UNDRR 2022a). At current trends, the number of disasters per year globally may increase from about 400 in 2015 to 560 by 2030—a projected increase of 40% (UNDRR 2022b).

Regardless of the source of the information in various reports, there has been a rise in the frequency of climate-related extremes, including floods and landslides (hydrological), storms and fogs (meteorological), or the two combined (hydro-meteorological) as well as drought and wildfires (climatological). But this increase is not evident for earthquakes and volcanoes (geophysical), an indication that the increase could be climate related (Fig. 2.1). This difference in the trends of hydro-meteorological and climatic events on the one side and geophysical ones on the other, in the sources used here, is a powerful and intuitive signal of the nature of climate change.

An analysis of the period since 1900 shows an increase in record-breaking rainfall events since around 1980. While the number of extreme rainfall events prior to 1980 was associated with natural multi-decadal variability, the spike in extreme events after 1980 is consistent with rising temperatures. Cyclones form when warm ocean waters, brought about

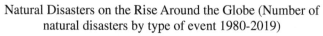
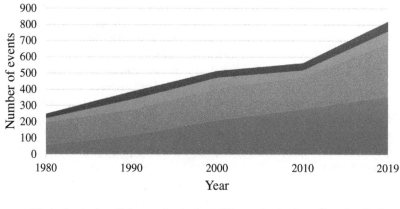

Fig. 2.1 Hazards of nature on the rise (*Source* Buchholz 2020; Munich RE 2022)

by global warming, create warm moist air. Each degree Celsius rise in sea surface temperatures is likely to double the frequencies of hurricanes (Chapter 6). The sea surface temperature of the Pacific Warm Pool region was at its highest when Typhoon Haiyan was formed in the Philippines in 2013.

As surface temperatures are fast exceeding previous records, 2022 was on track for seeing the deadliest heatwaves, droughts, and forest fires. The world's rivers are drying up at a fast clip (Croker et al. 2022). Along with the warming, sea levels have registered a steady rise worldwide. Atmospheric thermodynamics explain that warmer atmospheres have larger saturation vapour content, and increases in near-surface temperatures raise precipitation. Warming air also raises sea levels to dangerous levels and flooding lowers the ground level through underground inundation. One such case is Miami which is less than 2 metres above sea level (Lemonick 2012).

The summer of 2022 featured "off the chart" events to the point that once-in-100-year disasters seemed part of everyday life (Lopez 2022). Among disasters that fell into this category are the floods that submerged more than a third of Pakistan, killing more than 1,700 people, and costing an estimated US$30 billion in damages and economic losses according to the World Bank; heatwaves and wildfires that scorched the US and Europe; floods that hammered parts of the US; a heatwave and drought that dried up China's rivers; and epic rainfall just before summer, causing floods and mudslides in South Africa, killing at least 450 people.

The link between GHG emissions in the air, warming of the seas, and more intense natural disasters is powerful (see Chapter 6 also), and yet this connection is not recognised widely (Thomas 2017). The warming of the seas contributes to greater energy that adds force to the winds and greater moisture that makes for heavier rainfall. The airflows change as well. Warmer air also melts ice caps and glaciers while warmer waters raise sea levels, constituting conditions for perfect storms. Scientists have made clear these links in excruciating detail, and the policy implications for the urgency to slash discharges are crystal clear, but either for their inconvenience or for their complexity, the emissions-warming-disaster link is not being acted upon. Because weather catastrophes are in the natural world, it is easy to attribute them as acts of God—to use that insurance industry term—even though the source of their increasing intensity and frequency is human-made climate change.

Responding to climate change is unquestionably slowed by certain governments, glaringly, for example, the US under former President Donald Trump, and Brazil under former President Jair Bolsonaro denying the role of human-induced climate change, and worse, taking measures that aggravate the problem. The views of climate deniers may have cleverly evolved, from raising doubts on the presence of climate change to questioning whether it is natural or human-induced, to a more sophisticated argument on the extent of the impact and the effectiveness of remedies. For example, they call for lifecycle analysis of biofuels, whereas no such lifecycle analysis is demanded for the oil and gas industry.

It is sometimes argued that deaths from disasters have fallen and that is an indication that the gravity of the problem is lessening. Weather-related disasters show the biggest increase in incidence over the past 50 years. They are unmistakably causing more dislocation and greater damage, while causing fewer deaths (WMO 2021). Most of the deaths are taking place in developing countries. Just as life expectancy is up in general, the decline in mortality from these disasters reflects improvements in income, health, and infrastructure, as well as specific measures in disaster management like early warning and evacuation. But they do not suggest that disasters are down, or that they are less damaging.

Damages from hazards of nature have risen notably during 1990–2020, especially for floods and storms (UNDRR 2022a). A record number of floods, hurricanes, and wildfires intensified by global warming cost an estimated US$210 billion globally in 2020 (Newburger 2021). In July 2022, extreme heat prompted authorities to declare the heatwave a national emergency across the United Kingdom (UK), with Wales experiencing a record-high temperature of 37.1°C on 18 July, at a high cost to people's health (BBC News 2022). The insured losses in the US alone from Hurricane Ian that hammered the southwestern coast, according to S&P Global, a ratings company, will run to US$30–40 billion (S&P Global Ratings 2022). Total losses in this case would be far higher; and there were an estimated 18 billion-dollar events in the US alone. Future losses from environmental disasters will increasingly hurt economic growth, poverty reduction, and income distribution, as signalled in Chapter 5.

The Nature of Risk

Risk is typically seen as a jolt to the system and that dislocation is often considered as exogenous, out of the ordinary, and beyond control. But it is crucial to recognise that there are varying degrees of unexpectedness and of society's ability to deal with them as well as influence them. These features vary across geographical locations and can change over time depending on policies and investments.

Different Perils, Different Places

The world today faces several exogenous (or externally-imposed) and endogenous (internally-determined) shocks. The global climate crisis is powerfully affecting business operations, livelihoods, and geopolitical landscapes. The incidence of natural disasters varies in its frequency and across locations, for example, those in Indonesia and Vietnam rank high and in Laos and Singapore, rank low. Natural disasters, a perennial threat to lives and livelihoods in many regions of the world, have long been thought of as exogenous shockwaves that are outside of one's policy control. It is not uncommon to hear that a Category 4 hurricane cannot be avoided, but what matters is how well society recovers from it.

But the evidence on the nature of climate disasters suggests otherwise. The severity and frequency of hazards of nature are responsive to mitigation, while their impacts on the ground are influenced by adaptation. When human-induced global warming is added to the equation, the characteristics of disasters change, becoming at least in part endogenous and having elements of predictability and policy influence. Certain types of external shocks may prompt societies to change course, reform policies, and bounce back stronger.

An initial disturbance can work through the system differently depending on how preparedness modifies its trajectory. In a highly globalized world economy, macroeconomic risks are highly transmittable. The macroeconomy of a country, comprising households, businesses, and the government, responds to its fiscal and monetary policies, trade, and financial flows as well as political trends. What happens in a country is highly influenced by the behaviour of others, including partners in trade and investments. The cross-country impacts can be seen inter alia in exchange rates, interest rates, and prices, which are also variables a country uses to manage its macroeconomy.

Infections that cross borders are tangible manifestations of cross-country health impacts. COVID-19 was mostly an exogenous shock, but its consequences depended on a country's readiness to deal with the pandemic and to take further measures, such as contact tracing, mask mandates, and lockdowns, to curb the spread of the virus and lower infections. Countries deployed new approaches to respond to the public health crisis with shifting characteristics, resulting in both successes and failures. Two interlinked factors seem to have been key to effective management of the pandemic in East Asia's experience: first, the readiness and speed on the part of governments to adjust and modify response strategies according to changing circumstances; and second, their ability to maintain societal trust through principled decisions, consensus building, and effective communication.

Geopolitical conflicts are another example that can seem exogenous but are influenced by past investments and current efforts. For example, a minor incident in contested areas like the South China Sea can lead to vastly differing degrees of aggression and confrontation in the immediate region and beyond. Russia's 2022 invasion of Ukraine has multiple resulting paths, but a significant factor in shaping outcomes would be Ukraine's political will and investments in resilience building that contribute to the ability to fight back (The Hub 2022). Geopolitical conflicts can affect food security and increase food prices, as was the case in 2022. Many developing countries are under severe stress from debt, especially with interest rate increases, which interact with rising food prices (Landers and Aboneaaj 2022). Governments have run out of fiscal space to provide safety nets in the face of inflationary pressures. A widespread debt crisis could only exacerbate global trade and supply chains. The social ramifications are severe, especially given severe malnutrition and hunger in many developing economies. A report from the UN shows the damaging impact from the confluence of conflict, COVID-19, and global warming, especially on women's and children's health (WHO 2022).

Role of Policy

Natural disasters have long been considered intermittent and uncontrollable shockwaves. But because extreme weather events are becoming more frequent, they are no longer one-off events. Climate change is the result of actions taken over time, and the emerging scenarios depend on what has already been done, while efforts need to be made to change the trajectory over the next years. Scientific evidence points to human causes of global warming (IPCC 2021). This connection between climate change and the extremity of hazards of nature is at the centre stage of global events, potentially providing the most tangible and proximate rationale for the urgency of anti-pollution policies (Thomas 2017).

In this context, carbon emission is a common unit used to express the different types of GHGs that contribute to climate change (see Box 2.1). Carbon dioxide is also the most understood link, through the greenhouse effect, that raises sea level temperatures—one that might first have been measured by Swedish scientist Svante Arrhenius in 1896.

Box 2.1 Greenhouse Gases and Carbon Emissions
GHGs absorb and emit energy, causing the greenhouse effect. The primary GHGs in the atmosphere are the major long-lived components: water vapour (H_2O), carbon dioxide (CO_2), methane (CH_4), nitrous oxide (N_2O), chlorofluorocarbons (CFCs), carbon tetrachloride (CCl_4),

hydrofluorocarbons (HFCs), perfluorocarbons (PFCs), sulfur hexafluoride (SF_6), and nitrogen trifluoride (NF_3). CO_2 makes up some 76% of GHGs, while CH_4, heavily from agriculture and livestock, but also coal mining, oil and gas systems and landfills, is about 16%. Water vapour is an abundant GHG but also a consequence of global warming. Ozone is technically a GHG, but whether it is helpful or harmful depends on where in the atmosphere it is found.

GHG concentrations are measured in parts per million, parts per billion, and even parts per trillion. One part per million is equivalent to one drop of water diluted into about 13 gallons of liquid (roughly the fuel tank of a compact car). Larger emissions lead to higher concentrations in the atmosphere. "Carbon dioxide equivalent" is a term for describing different GHGs in a common unit. For any quantity and type of GHG, carbon dioxide equivalence signifies the amount of CO_2 which would have the equivalent global warming impact.

GHG molecules in the atmosphere absorb light, preventing some of it from escaping the Earth. They act like a blanket, trapping heat and warming the surface rather than allowing the heat to escape into space. Each of these gases can remain in the atmosphere for different amounts of time, ranging from a few years to thousands of years. Some gases are more harmful than others in making the planet warmer and thickening the Earth's blanket.

Carbon discharges come at a heavy cost to society, and there have been various efforts to put a price on them. A recent estimate finds that an additional ton of carbon dioxide in the air costs society US$185, which is more than three times the figure of US$51 the US government has used (Rennert et al. 2022). In 2020, New York state adopted a price of US$79–US$125 for its programmes. The figure of US$185 is the result of accounting for climate-related mortality, damages from weather disasters, and the value of crop failure.

Figure 2.2 shows, as do other data sources, that carbon emissions and temperature changes are highly and positively correlated, and this relation spans a lengthy period. This positive correlation between temperature change and emissions became stronger post-1970. Note that the temperature change in this figure is obtained by subtracting the first data point in the chart ($-0.16\ °C$) from the latest data point ($0.85\ °C$). The July 2022 temperature was $1.15\ °C$ higher than the 1880–1920 average. What is more, the underlying CO_2 concentration in the atmosphere is still headed the wrong way. The August 2022 number was 417.19 parts per million by volume (ppmv) compared to 414.47 ppmv a year before (Earth's CO_2 Homepage; see Chapter 5).

There have been efforts to predict the occurrence and intensity of shocks, creating more time for preparation and mitigation measures (Webster and Jian 2011). This can reduce the eventual severity of economic, societal, and life costs. Regions and localities that face more frequent risks of weather events, such as South Asia and East Asia and their states, provinces and localities, will benefit from efforts to predict extreme events. Recurring systematic risks can be reduced with calculated anticipation of the time and severity of the adverse event (Courtnell 2020).

Webster and Jian (2011) found that forecasts in the Brahmaputra and Brahmaputra–Ganges areas helped reduce damage risks, based on criteria including economic savings, drought risk index, and freshwater availability. The impact of floods on the global population is significant, as the Brahmaputra–Ganges areas house 14% of the global population. Flood forecasts were able to provide communities with information on the timing, duration, and intensity of floods, hence buying time to make

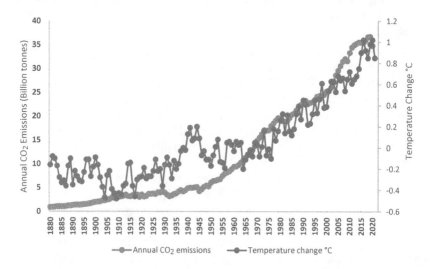

Fig. 2.2 Carbon emissions and temperatures (*Source* Andrew and Peters 2021; Global Carbon Project 2021; NASA 2022; NOAA 2022)

the necessary preparations or evacuations. Through advanced planning and actions by communities and countries, the economic and life risks associated with these events were reduced. Given the environmental deterioration that has been taking place, all regions will become increasingly vulnerable. In this regard, Chapter 5 will suggest that the past may not be a sufficient guide for the future as the Philippines in 2013, Syria in 2021, and Pakistan in 2022 found that no amount of planning would have prepared them for the hurricanes, droughts, and floods they respectively faced.

Natural hazards do not cause extreme disasters all by themselves—disasters happen when a hazard event interacts with an exposed inhabited area, particularly areas with vulnerable populations, economic assets, and an environment that results in disasters. Risk, the metric of a disaster, combines the probability of hazardous events taking place and the potential impact of these events on lives and livelihoods, health, ecosystems, economies, societies, cultures, services, and infrastructure.

The evidence on the human-made nature of climate disasters makes them potentially responsive to mitigation (Royal Society 2014). Mitigation involves lessening adverse impacts by preventing or reducing the intensity of hazards. For climate change, mitigation is about reducing the sources of GHGs; for example, phasing out coal plants and replacing them with renewable energies, or increasing the storage of carbon by protecting forests. That in turn influences the energy intensity of storms or the precipitation underlying rainfall and floods and the heat that contributes to droughts and wildfires.

Adaptation, as the term suggests, is taking steps to adjust to adverse events, including measures to reduce damage. Adaptation is understood as the process of adjusting to the effects, for example, by building coastal defences, such as seawalls or improving drainage systems. Adjusting to climate change is essential to survival especially given the momentous changes already underway based on the stock of carbon that has accumulated in the atmosphere and is envisaged to stay for decades regardless of mitigation efforts (see UNFCCC 2022).

Some steps help both mitigation and adaptation. For example, protecting mangrove forests along the coastlines, or stemming deforestation, provides prevention by functioning as carbon sinks, and protecting people living in the line of storms and floods. Some measures can serve as mitigation in one context and adaptation in another. An example is the construction of "sponge cities", which in coastal regions increases resilience to floods and waterlogging (i.e., adaptation) whereas in inland areas, they increase the carbon sink (i.e., mitigation). Another is forest fire detection, which can prevent fires when they are started by human action, or can help control them better when forest fires occur naturally.

Timing Can Be Everything

One attribute of risk is the time frame in which a risk presents itself. The variables to watch in the case of climate change are the pace of increase in effluents and their effect on global temperatures, somewhat analogous to the rate of increase in the incidence of infections and hospitalisation in the case of the COVID-19 pandemic. How the impact on the system is spread out over time, or not, fundamentally affects the ability of systems to function and even bounce back.

In Fig. 2.3, a challenging element of both climate change and the pandemic is the capacity to stretch the timeframe of the shock, limiting it to more manageable levels, and thus allowing governments to respond. It makes a big difference if the temperature impact or the number of infections is concentrated over a short period or spread out over time. For global warming, the extremity of disasters and the ability to cope are shaped by slowing the pace of the warming that drives the temperature to a given level as in a depiction by David J. Hayes at NYU Energy & Environmental Impact Center. In the case of the pandemic, it is the difference between a manageable caseload and the intensive care units being unable to cope.

For both the pandemic and global warming, the reported situation could be an underestimate of the reality. For the health system, a key indicator is the Intensive Care Unit (ICU) space but that may be only a fraction of the very ill. Nevertheless, a country adopts measures commensurate with the extent of ICU capacity. In the climate system, even a net zero target for 2050 could be inadequate to keep temperature rise to below 2 °C, if not 1.5 °C, and prevent harsh outcomes, especially as a string of damages would already have been locked in. For instance, in the case of the ozone layer, even when CFCs were prohibited, it took many decades for the ozone hole to partially recover. In the case of global warming too, the climate system might need to overshoot the current targets to ensure returning to any semblance of normalcy.

Society's relatively swift response to COVID-19 unlike the slow and often counter-productive reaction to global warming raises fundamental questions. Is the priority given to a problem related to perceptions of the

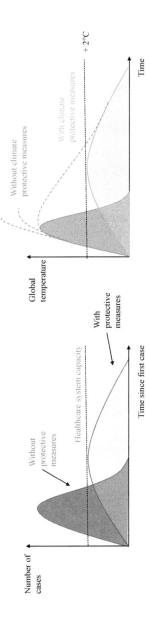

Fig. 2.3 Comparing risks of climate change and the pandemic (*Source* Based on Gourinchas 2020; Verkooijen 2020)

imminence of outcomes? Are risks that are mostly outside one's control, such as pandemics, valued more than ones that can be influenced by own actions like a financial crisis? Is attention more easily directed to more immediate and modest issues that are manageable, while postponing the focus on the slower onset of intractable ones, even if they have ultimately deadlier consequences?

THE SIGNIFICANCE OF RESILIENCE

Broadly speaking, resilience refers to the ability of a system to sustain or restore its functionality and performance following a shock or a change in the workings of the system. Management of resilience benefits from the quality and timeliness of risk assessments. Risk-resilience analysis extends beyond quantitative risk assessments using objective data to determine the value of assets and to measure the probability of loss under alternative scenarios. It also includes qualitative considerations highlighting uncertainties and needed responses.

Facets of Resilience

The World Bank usefully sees risk management as comprising interlinked components that span preparation to coping (Fig. 2.4). Knowledge and knowledge sharing are key to understanding risks and reducing the uncertainty surrounding them. Insurance can help to transfer or distribute risks in ways that are welfare enhancing. Coping is a big part of recovering from the downswing and finding profitable avenues. Protection signifies ways and means for lowering the probability of losses and improving opportunities for gains.

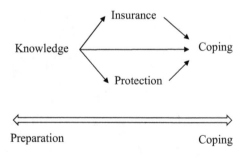

Fig. 2.4 The prevention–response axis of disasters (*Source* Taken from World Bank 2013)

Resilience building can take different forms. One is to avoid risk by choosing not to undertake some activities that are perceived to be highly risky. Another is to transfer risk to third parties through insurance, hedging, and outsourcing, an avenue that is underutilised in many developing countries. There is also the option to accept risk in some instances, recognising that undertaking certain activities entails risks but nevertheless generates an expected outcome that justifies it. Taking steps to mitigate risk by taking preventive measures is the other way forward, a primary focus of this book.

Reports and eye-witness accounts of high-risk episodes are filled with stories of people's remarkable resilience in the face of challenges. But it is worth drawing attention to COVID-19, which has met very different degrees of responses depending on the setting and the country. The type of damage during a pandemic compared with weather disasters is of course different. But one distinction this book is making is that if the jolt is beyond one's control, it makes eminent sense to credit people for their resilience and the preparations they make for the future. But if the shock is endogenous, it is imperative to also call out the sources of the problem as part of preventive measures to avert worse outcomes in the future.

Past performance or indicators of resilience may not guarantee future success. The Global Health Security Index 2019 ranked the US first among 195 nations in the strength of national health security just prior to the onset of COVID-19 (Cameron et al. 2019). But Bloomberg's COVID Resilience Ranking (2021) had the US 36th out of 53 countries considering the evidence on infections, hospitalisations, and the like. Countries have often been repeatedly stymied by health crises. But some have found ways to turn devastating situations around, whether through science, social cohesion, or simply learning from the past.

Although no top performer sustained success over time in the Bloomberg rankings since 2020, Norway, Denmark, Finland, the United Arab Emirates, Canada, South Korea, and Switzerland were notable in their success early on in achieving high vaccination rates, battling new COVID-19 variants, and in economic recovery. Strong healthcare safety nets and societal cohesion are common factors that benefited countries at every stage of the pandemic. Faith in government and people's willingness to follow rules helped countries contain the virus. At the end of 2020, some were notably hard hit in infection numbers and vaccination rates, for example, Argentina, Iran, Mexico, Brazil, Peru, Poland, Nigeria,

Pakistan, and South Africa (see also Tsou et al. 2022). China's policies since then stand out for their low death rates but huge disruptions and a heavy hand of government.

Individuals and Society

Policies for building resilience must also consider the benefits to individuals as well as to the larger society. Some measures—for example, the policies implemented by countries to control COVID-19—are win–win in the sense that the principal stakeholders win. These policies are expected to gain widespread acceptance. But others could be win-lose at the individual level where some parties are clear winners, but others are disadvantaged by policies. What makes this interesting from a policy point of view is when, despite some losers, a net-win or a societal gain on balance happens.

Improving energy efficiency is a win–win approach using policies and behaviour change (Petrie and Thomas 2013). In other areas, costs are involved making for win–lose situations, in that some gain and some may lose. Yet, they may represent net-win opportunities, in the sense that the costs to some are more than offset by societal gains. Removing fossil fuel subsidies would mean losses to some, but it improves societal well-being by reducing GHGs (van den Berg and Cando-Noordhuizen 2017). The question is what it would take for society to accept net-win policies.

CLIMATE RISK AND RESILIENCE

Climate change presents extreme versions of net-win outcomes by bringing climate and sustainability together (Munasinghe 1999). But countries are still reluctant to pursue sustainable development. The time dimension is clearly a complicating factor in that the win-lose picture of the short-term may only be matched by a net-win situation after some time. Economic actors may be unwilling to take short-term losses for promised gains—often in terms of avoided losses—in the longer term.

Climate change exemplifies the point that resilience building can modify the probability of shocks to the system, at least in part, thus underscoring the possibility of influencing risks through policy. Climate change can be seen to generate a series of shocks, such as floods and storms, that can be contained through policy action. Adaptation is essential given the inevitability of large systemic changes already afoot. But mitigation is urgently needed to lessen their frequency and severity of the shocks, a theme we return to in Chapter 8 in the context of policy responses.

Reaching and sticking to global agreements among all the nations are fraught with difficulties. Given this tough background, the 2015 Paris Agreement, the architect of which was the then Executive Secretary of the United Nations Framework Convention on Climate Change (UNFCCC), Christiana Figueres, was a landmark event to significantly reduce climate risks and impacts. The intention was to work within a framework whereby countries work towards keeping global warming "well below" 2 °C, and to "make efforts" to keep it to 1.5 °C. The goals are highly significant. There is also evidence that establishing a policy framework for low-carbon growth makes a difference to actions by the government and businesses to invest in green and low-carbon projects (Zheng 2020). This finding has significant implications for the potential of market policies to underpin green investments.

That said, the steps promised and progress in implementation thus far are very disappointing. The major concerns over the Paris Agreement are the adequacy of the commitments and the degree of their enforceability. In the 2021 COP26 Glasgow Climate Pact, countries renewed their commitments, though in varying degrees, to more intensive efforts. These included transitioning towards carbon neutrality, reducing reliance on energy-inefficient fossil fuels, and accumulating US$100 billion in climate financing annually (Tan and Fogarty 2021). But one estimate is that for the 1.5 °C goal, GHGs must decline by 45% by 2030 compared with 2019 levels, but even with the fulfilment of the Paris commitments (which is far from being the case), emissions will fall only 7.5% by 2030 (Menon 2022).

In general, commitments and their fulfilment might be seen in the context of the assessment of the IPCC that global emissions need to reach net zero by about mid-century to give a reasonable chance of limiting warming to 1.5 °C (IPCC 2018). Against such a benchmark, Climate Action Tracker (2022) continues to rate many countries' efforts to fulfil pledges thus far as insufficient, or worse, to indicate a need for much stronger commitments.

Box 2.2 Carbon Neutrality and the Paris Agreement

Carbon[1] neutral means striking a balance between emitting carbon and offsetting it by absorbing carbon, thereby leaving no net additions in the atmosphere. Net zero carbon means making changes to reduce carbon emissions to the lowest amount—and offsetting only as a last resort. There are several examples of efforts towards carbon neutrality. For one, investment in renewable energy will help reduce reliance on fossil fuels that are the primary cause of carbon emissions. Achieving greater energy efficiency and making lifestyle changes to reduce the amount of energy consumed is another.

These outcomes call on societies to make changes in lifestyles. The changes can be nudged by pricing carbon emissions in countries, which can take the form of putting a tax on emissions or establishing a market for carbon emissions. It can also include imposing a border carbon tariff on imports, as 20–30% of the emissions are estimated to be contained in goods and services that are internationally traded (WTO 2021). Accounting for this carbon content would be important as the divergence between consumption and production-based emissions can be large. For example, in place of a 3% increase in production-based emissions since 1990, the US would show a 14% increase in consumption-based emissions (Ritchie 2019). Changing policies based on such findings require a commitment on the part of governments and businesses for policy reform and implementation.

More than 190 countries signed the Paris Agreement, an international treaty on climate change that was reached at the COP21 in 2015. As noted, the Agreement aims to keep a rise in global temperatures to below 2 °C, if not 1.5 °C, relative to pre-industrial levels, by the end of this century. Since then, a growing number of countries have been signing onto voluntary commitments of net zero emissions by 2045, 2050, 2060, or 2070. There is also a growing awareness that these commitments and their implementation are too little, too late. Furthermore, the inclination to think of climate change as a distant problem is unhelpful. The damages and costs are already mounting, and what is more, the stock of accumulating effluents delays the impact of actions taken. It is increasingly becoming apparent that what is needed is not just carbon neutrality, and much earlier than previously thought, but objectives for achieving carbon-negative outcomes. Indeed, the critical question concerns the more immediate cuts in effluents, and the resulting trajectory countries will likely have, say by 2030.

Conclusions

A principal observation is the rise in the number of extreme climatic disasters over the past 50 years, resulting in a rise in damages related to these events. Historically, natural disasters were considered exogenous shocks. But now that global warming is added to the equation, their characteristics are understood to have changed, becoming—at least in part—endogenous. As they become more frequent, they can no longer be considered stand-alone events, but rather as part of a scenario that can be expected to worsen.

Risk and resilience go together. Risk assessments inform the design and implementation of resilience management. Resilience building involves avoiding risks, transferring risks to third parties, accepting risks, or mitigating risks through preventive measures. Global responses to COVID-19 have shown that an initial disturbance can work through the system differently, depending on how preparedness modifies its trajectory. The challenge of both climate change and pandemics can be seen as the capacity to stretch the shock's timeframe. For climate change, the extremity of disasters and the ability to cope with them are shaped by slowing the pace of the warming.

The Paris Agreement provides some hope that climate measures can indeed be taken. But countries need to strengthen their commitments and deliver on them. Achieving carbon neutrality calls for a vast cut in their reliance on fossil fuels, and mobilisation of large financing to enable a low-carbon transition. The search for new solutions such as commercially viable green hydrogen or carbon capture needs to be supported and sped up. These are the necessary conditions and the initial reforms needed in a difficult journey to tackle climate change.

Note

1. While carbon dioxide is one of the principal GHG emissions causing global warming, the term "carbon" is often used to refer to all GHGs.

Bibliography

Andrew, Robbie M., and Glen P. Peters. 2021. "The Global Carbon Project's Fossil CO_2 Emissions Dataset." *Zenodo.* https://doi.org/10.5281/zenodo.5569235.

Anstey, Chris. 2022. "Summers Says Global Market Risk Is Building Like in August 2007—Bloomberg." *Bloomberg Asia Edition*, September 29, 2022, sec. Markets. https://www.bloomberg.com/news/articles/2022-09-29/summers-says-global-market-risk-is-building-like-in-august-2007.

Berg, Rob D. van den, and Lee Cando-Noordhuizen. 2017. "Action on Climate Change: What Does It Mean and Where Does It Lead To?" In *Evaluating Climate Change Action for Sustainable Development*, edited by Juha I. Uitto, Jyotsna Puri, and Rob D. van den Berg, 13–34. Cham: Springer. https://doi.org/10.1007/978-3-319-43702-6_2.

Bloomberg News. 2021. "The Winners and Losers From a Year of Ranking COVID Resilience." *Bloomberg.Com*, November 23, 2021. https://www.bloomberg.com/news/features/2021-11-23/the-winners-and-losers-from-a-year-of-ranking-COVID-resilience.

British Broadcasting Corporation. 2022. "Heatwave: Wales' Hottest Day as Temperature Hits 37.1C." *BBC News*, July 18, 2022, sec. Wales. https://www.bbc.com/news/uk-wales-62154870.

Buchholz, Katharina. 2020. "Natural Disasters on the Rise around the Globe." *Statista*. August 25. https://www.statista.com/chart/22686/number-of-natural-disasters-globally/.

Cameron, Elizabeth E., Jennifer B. Nuzzo, Jessica A. Bell, Michelle Nalabandian, John O'Brien, League Avery, Sanjana Ravi, et al. 2019. "Global Health Security Index: Building Collective Action and Accountability." *GHS Index*, October, 324.

Climate Action Tracker. 2022. "State of Climate Action." https://climateactiontracker.org/.

Courtnell, Jane. 2020. "Systematic Risk: The Difference and Its Relation to Systemic Risk." https://www.process.st/systemic-risk/.

Croker, Natalie, Renee Rigdon, Judson Jones, Carlotta Dotto, and Angela Dewan. 2022. "The World's Rivers Are Drying up in Drought and Heat. Here's How 6 Look from Space." https://edition.cnn.com/2022/08/20/world/rivers-lakes-drying-up-drought-climate-cmd-intl/index.html.

Curry, Judith. 2023. *Climate Uncertainty and Risk*. Anthem Press. July.

Earth's CO_2 Homepage, https://www.CO2.earth/.

Global Carbon Project. 2021. "Supplemental Data of Global Carbon Budget 2021 (Version 1.0) [Data Set]." ICOS. https://www.icos-cp.eu/science-and-impact/global-carbon-budget/2021.

Gourinchas, Pierre-Olivier. 2020. "Flattening the Pandemic and Recession Curves." *The Unassuming Economist*, March 15, 2020. https://unassumingeconomist.com/2020/03/flattening-the-pandemic-and-recession-curves/.

Iberdrola. 2019. "Which Countries Are Most Threatened by and Vulnerable to Climate Change?" *Iberdrola*. https://www.iberdrola.com/sustainability/top-countries-most-affected-by-climate-change.

Institute for Economics and Peace. 2020. "Ecological Threat Register 2020: Understanding Ecological Threats, Resilience and Peace." Vision of Humanity.
Intergovernmental Panel for Climate Change. 2018. "Global Warming of 1.5 °C." https://www.ipcc.ch/sr15/.
Intergovernmental Panel on Climate Change. 2021. Sixth Assessment Report. *AR6*. September. https://www.ipcc.ch/assessment-report/ar6/.
Intergovernmental Panel on Climate Change. 2022. "Chapter 4: Sea Level Rise and Implications for Low-Lying Islands, Coasts and Communities — Special Report on the Ocean and Cryosphere in a Changing Climate." https://www.ipcc.ch/srocc/chapter/chapter-4-sea-level-rise-and-implications-for-low-lying-islands-coasts-and-communities/.
International Monetary Fund. 2022. "Global Financial Stability Report." https://www.imf.org/en/Publications/GFSR.
Landers, Clemens and Rakan Aboneaaj. 2022. "How the Global Debt Crisis Could Make the Hunger Crisis Worse". Blogpost, October 17. Center for Global Development. https://www.cgdev.org/blog/how-global-debt-crisis-could-make-hunger-crisis-worse.
Lemonick, Michael D. 2012. "The Future Is Now for Sea Level Rise in South Florida." https://www.climatecentral.org/news/the-future-is-now-for-sea-level-rise-in-south-florida.
Levitan, Dave. 2022. "The Planet Is on Fire — but 2022 Won't Crack the Grim Top 5 List of Warmest Years on Record." https://www.grid.news/story/climate/2022/07/20/the-planet-is-on-fire-but-2022-wont-crack-the-grim-top-5-list-of-warmest-years-on-record/.
Lopez, German. 2022. "A Summer of Climate Disasters—The New York Times." *The New York Times*, September 7, 2022. https://www.nytimes.com/2022/09/07/briefing/climate-change-heat-waves-us-europe.html.
Menon, Ravi. 2022. "'What Does It Take to Get to Net Zero'—Keynote Speech by Mr Ravi Menon, Managing Director, Monetary Authority of Singapore, at the Economic Society of Singapore Annual Dinner 2022 on 17 August 2022." https://www.mas.gov.sg/news/speeches/2022/what-does-it-take-to-get-to-net-zero-keynote-speech-by-mr-ravi-menon-managing-director-monetary-authority-of-singapore-at-the-economic-society-of-singapore-annual-dinner-2022-on-17-august-2022.
Munich RE. 2022. "Natural Disaster Risks: Losses Are Trending Upwards | Munich Re." https://www.munichre.com/en/risks/natural-disasters-losses-are-trending-upwards.html.
Munasinghe, M. 1999. "Sustainomics, Sustainable Development and Climate Change." *International Journal of Space Structures*, 14 (4): 393–414. https://doi.org/10.1260/0266351991494920.

National Aeronautics and Space Administration. 2022. "Global Surface Temperature | NASA Global Climate Change." Climate Change: Vital Signs of the Planet. https://climate.nasa.gov/vital-signs/global-temperature.

National Oceanic and Atmospheric Administration. 2022. "Climate Change: Atmospheric Carbon Dioxide." June 23. https://www.climate.gov/news-features/understanding-climate/climate-change-atmospheric-carbon-dioxide#:~:text=Without%20carbon%20dioxide%2C%20Earth's%20natural,causing%20global%20temperature%20to%20rise.

Newburger, Emma. 2021. "Disasters Caused $210 Billion in Damage in 2020, Showing Growing Cost of Climate Change." *CNBC*, January 7, 2021. https://www.cnbc.com/2021/01/07/climate-change-disasters-cause-210-billion-in-damage-in-2020.html.

Petri, Peter, and Vinod Thomas. 2013. "Asia's Coming Shift." *The Magazine of International Economic Policy*. http://www.international-economy.com/TIE_Su13_PetriThomas.pdf.

Pulwarty, Roger, Loretta Hiebert-Girardet, Ricardo Mena Speck, Erica Allis, Cyrille Honoré, and Johan Stander. 2022. "Risk to Resilience: Climate Change, Disasters and the WMO-UNDRR Centre of Excellence." https://public.wmo.int/en/resources/bulletin/risk-resilience-climate-change-disasters-and-wmo-undrr-centre-of-excellence.

Rennert, K., F. Errickson, B.C. Prest, et al. 2022. "Comprehensive Evidence Implies a Higher Social Cost of CO_2". *Nature*. https://doi.org/10.1038/s41586-022-05224-9.

Ritchie, Hannah. 2019. "How do CO_2 emissions compare when we adjust for trade?" Our World in Data, October 7. https://ourworldindata.org/consumption-based-co2.

S&P Global Ratings. 2022. "Insured Losses From Hurricane Ian Will Likely Be Substantial But Manageable." https://www.spglobal.com/ratings/en/research/articles/220930-insured-losses-from-hurricane-ian-will-likely-be-substantial-but-manageable-12516796.

Tan, Audrey, and David Fogarty. 2021. "Carbon Copy: The State of Play on the Six Key Issues at UN Climate Conference COP26." *The Straits Times*, November 2021. https://www.straitstimes.com/world/carbon-copy-the-state-of-play-on-the-six-key-issues-at-un-climate-conference-cop26.

The Hub. 2022. "Ukraine's Resilience, Russia's Miscalculation." *The Hub*. https://hub.jhu.edu/2022/03/02/sais-panel-ukraine-russia-conflict/.

The Royal Society. 2014. "Climate Change: Evidence and Causes|Royal Society." https://royalsociety.org/topics-policy/projects/climate-change-evidence-causes/basics-of-climate-change/.

The World Bank. 2013. *World Development Report 2014: Risk and Opportunity—Managing Risk for Development*. Washington, DC: World Bank. https://openknowledge.worldbank.org/handle/10986/16092 License: CC BY 3.0 IGO.

Thomas, Vinod. 2017. *Climate Change and Natural Disasters: Transforming Economics and Policies for a Sustainable Future.* 1st ed. New Brunswick: Transaction Publishers. https://doi.org/10.4324/9781315081045.
Tsou, H.H., S.C. Kuo, Y.H. Lin, et al. 2022. "A Comprehensive Evaluation of COVID-19 Policies and Outcomes in 50 Countries and Territories. "*Science and Reports* 12: 8802. https://doi.org/10.1038/s41598-022-12853-7.
United Nations Framework Convention on Climate Change. 2022. "What Do Adaptation to Climate Change and Climate Resilience Mean?|UNFCCC." https://unfccc.int/topics/adaptation-and-resilience/the-big-picture/what-do-adaptation-to-climate-change-and-climate-resilience-mean.
United Nations Office for Disaster Risk Reduction. 2022a. "GAR2022a: Our World at Risk." https://www.undrr.org/gar2022a-our-world-risk.
United Nations Office for Disaster Risk Reduction. 2022b. Global Assessment Report on Disaster Risk Reduction. https://www.undrr.org/publication/global-assessment-report-disaster-risk-reduction-2022b.
Verkooijen, Patrick. 2020. "Flattening the Climate Curve in the Post-COVID World." World Economic Forum, April 17, 2020. https://www.weforum.org/agenda/2020/04/flattening-the-climate-curve-in-the-post-covid-world/.
Webster, Peter J., and Jun Jian. 2011. "Environmental Prediction, Risk Assessment and Extreme Events: Adaptation Strategies for the Developing World." *Philosophical Transactions. Series A, Mathematical, Physical, and Engineering Sciences* 369 (1956): 4768–97. https://doi.org/10.1098/rsta.2011.0160.
Wiedmann, Thomas, Arumina Malik, Glen Peters, Jacqueline Peel, and Xuemei Bai. 2022. "What Do the Findings of the IPCC Report Mean for Our World?" World Economic Forum, April 11, 2022. https://www.weforum.org/agenda/2022/04/ipcc-cut-global-emissions/.
World Economic Forum. 2022. "The Global Risks Report 2022 (17th Edition)." World Economic Forum. https://www3.weforum.org/docs/WEF_The_Global_Risks_Report_2022.pdf.
World Health Organization. 2022. "Every Woman Every Child 2022 Progress Report: Protect the Promise". https://www.who.int/news/item/18-10-2022-staggering-backsliding-across-women-s--children-s-and-adolescents--health-revealed-in-new-un-analysis.
World Meteorological Organization. 2021. *WMO Atlas of Mortality and Economic Losses from Weather, Climate and Water Extremes (1970–2019) (WMO-No. 1267).* Geneva: WMO.
World Trade Organization. 2021. "Trade and Climate Change. Information Brief No 4." https://www.wto.org/english/news_e/news21_e/clim_03nov21-4_e.pdf.

Zheng, Huanhuan. 2020. "Climate Policy and Sustainable Investments around the World", Lee Kuan Yew School of Public Policy. National University of Singapore. Working paper.

CHAPTER 3

Understanding Risk

If you thought the COVID pandemic was disruptive and deadly, climate change will be so much worse. A panellist at the United Nations Climate Summit 2021.

As climate hazards strike with increasing frequency, there is a growing awareness of something that is not only deadly but also more predictable. In the era of climate change, the occurrence of a once-in-a-100-year event is taking place, ominously, with far greater frequency. Elements of such a shift in outlook might be present in varying degrees for other risks, such as pandemics and food shortages. The significance of this change, from being rare but highly damaging to frequent and highly damaging, is that the bar for preparedness and response to these risks will be continuously rising.

With rising carbon emissions and persistent global warming, an increase in the incidence of extreme disasters is certain. An essential difference in predictability is the role of humankind rather than the wrath of nature in their incidence and extremity. The human contribution to these hazards—for example, through the destruction of the natural environment—puts into sharper focus the part that prevention must play in dealing with them. The more disasters owe to human activities, the more their impact can be addressed by development investments.

© The Author(s), under exclusive license to Springer Nature Singapore Pte Ltd. 2023
V. Thomas, *Risk and Resilience in the Era of Climate Change*, https://doi.org/10.1007/978-981-19-8621-5_3

The Global Risk Landscapes

The World Economic Forum's (WEF) report, *Global Risk Landscape 2021*, highlights different types of risks according to the likelihood of occurrence and their impact or damage (Fig. 3.1). Inequalities in healthcare, digitalisation, and education contribute to disproportionately higher risks to the lower-income strata. This assessment also illustrates the root trio of crises—health emergencies, climate change, and technology—that have pushed the world towards the brink of greater all-around risks (Bremmer 2022). The quadrant signifying high impact and high likelihood is of special policy interest (Fig 3.1).

Within this quadrant, the top three risks by likelihood fall within the environmental risk category—extreme weather, climate action failure, and human environmental damage. The top three risks by impact are infectious diseases, climate action failure, and biodiversity loss. The ranking of the top risks may not change much in a three-year span, but it does over 5–7 years.

Comparing World Economic Forum's *Global Risk Reports* over the years, one can discern changes in the ranking of perceived risks, with new ones emerging and old ones receding in importance. Since 2016, the top three in terms of likelihood were extreme weather events (each year), failure of climate actions (four out of six years), and natural disasters (each year). Cyberattacks and involuntary migration were in the top

Fig. 3.1 Perceptions of global risk (*Source* Based on World Economic Forum 2021)

three in some years. Extreme weather, climate change, and natural disasters were featured among the top risks in terms of impact in most years since 2016. Weapons of mass destruction, water crises, biodiversity loss, and infectious diseases were also prominent.

For cyberattacks and cybersecurity, the stakes are especially high for the operations of governments and businesses. The public and the private sectors are mindful that successful cyberattacks result in substantial damage to bottom lines, reputations, and productivity. Business enterprises are going to great lengths to guard against these attacks. Severe consequences are also being felt from personal data leaks and system hacks (McKinsey & Company 2020). Economy-wide effects of cyber insecurity can be devastating.

The assessment of risk is naturally and integrally accompanied by considering resilience, which is examined in the next chapter. Building resilience is an antidote to the presence of risk, and in the case of cyberattacks, it can readily be seen as influencing the chances and damages of these events. Here, resilience-building starts with individual awareness of cybersecurity risks and taking steps to guard against vulnerabilities. Box 3.1 describes some recent cyberattacks and efforts to build resilience.

Box 3.1 Risk and Resilience in Cybersecurity

Concerns about cybersecurity are worldwide and growing. They relate to the loss of confidentiality, integrity, or availability of data and information, and control over systems governing them. By endangering critical assets and sensitive information, they present financial and reputational harm to organisational operations. Application of technologies and controls for system protection is intended to avert cyberattacks and guard against system failures.

One example of resilience in cybersecurity is Maersk's response to the NotPetya malware cyberattack in 2017, which disrupted its operations by erasing its data and obliterating servers. Maersk's team of software engineers was able to mitigate the damage through reverse engineering. In the aftermath, resilience building initiatives included cybersecurity education for its employees, increasing investments in recruitment to build back operations, and strategies to detect and respond early (Ritchie 2019).

Contributing to this effort are software companies like Microsoft, whose Digital Crimes Unit works in tandem with governments and global businesses to thwart cyberattacks (Burt 2021). Other companies like IronNet make defence software available to organisations, alerting

> them to network threats instantaneously (Business Wire 2021). These collective exercises help entities guard against cyber risks, and through sufficient funding and stakeholder collaboration, improve individual entities' resilience to malicious action, thus reducing vulnerability to damages to livelihoods and security.

Interactions that Inform Policy

Understanding risk is also about recognising the multiple types of interlinkages involved. One relationship of great policy relevance is the common root causes of different categories of risk that accentuate actions needed to deal with them (Chivian and Bernstein 2008). For example, deforestation, driven by population pressure and illegal activities, not only causes loss of precious habitats and biodiversity but also can be germane to migration of species and spillover infections that can trigger pandemics. More sustainable management of flora, fauna, and livestock can be at the root of staving off the risk of certain infections as well as reducing GHGs and global warming.

Another linkage of considerable policy significance is that a category of risks can be transmitted from one setting to another, creating cross-border or cross-sectoral impacts. For instance, global warming manifested in sea level rises in certain locations also means declining crop yields, food insecurity, and trade dislocations elsewhere (Simpson and Trisos 2021). Multiple impacts across networks are a hallmark of cyber security risks. Risk assessments, to have a policy effect, need to account for interactions across multiple manifestations of a risk category.

Also, one type of risk can aggravate another, for example, climate risk heightening pandemic risk by destabilising the natural order of things. Several reviews of the COVID-19 experience have noted that those facing the risk of respiratory illnesses from air pollution are more likely to succumb to the pandemic, other factors being the same (Wu et al. 2020). There were also similar findings on the relation between air pollution and Severe Acute Respiratory Syndrome (SARS) (Kan et al. 2005). Similar observations have also been made on the relation between being exposed to a high degree of air pollution and the risk of smoking cigarettes.

A direct interaction is the effect of losses from a risk like climate change, or simultaneous risks say from climatic, economic, and health shocks, on the finances of a company or a local or national government. The direct hit of a risk like ecological destruction has often been underestimated. Measuring the impact of several events interacting with each other, while difficult to assess, is also important for being prepared (Ranger et al. 2021).

It is instructive to assess how the risk of a collapse of the ecosystem works its way through the supply chain to translate into the company's risks (Dasgupta 2021). Ecological risks can be shown to translate into financial risks at the level of business, and it is important for firms to make an adjustment to their expected value creation in terms of the losses in ecosystem services. It is possible to derive a risk-adjustment factor (that is between 0 and 1) that reflects the risk of ecological collapse, the ecosystem value of which should weigh on the enterprise's finances.

The term "systemic risk" is widely used in finance and medicine to denote eventualities associated with how approaches and systems work (Sillmann et al. 2022; UNDRR 2022). Again, the risk can denote a more favourable or less favourable outcome than normally anticipated. Its systemic nature derives from how the key components in a system, for example, health, infrastructure, and climate in the case of a pandemic, interact with each other. It is the human hand in shaping outcomes that renders the systemic nature of certain risks (Strader 2022). The present discussion of the environment and climate makes clear the systemic nature of disaster risks depending on how sustainable or not human actions are.

Risk and Uncertainty

It is important to recognise the distinction between *risk* and *uncertainty* (Barnett et al. 2022). In academia and the government, the discourse on climate policy makes multiple references to *risk*, usually while analysing pertinent bugbears in economics and finance. *Risk aversion* is a common term used in these discussions in areas such as where to invest resources for climate action and pandemic management. Borrowing a term from the world of finance, *risk premia* is also discussed in reference to understanding the differences in expected returns to financial, physical, and human investments.

Consideration of risk presumes knowledge of probabilities of outcomes, whereas uncertainty does not. Following the late economist Frank Knight, *uncertainty* can be seen as a broader term, while *risk* is more specific. Decision-making models based on risk aversion assume known probabilities, but not outcomes, such as in the case of outcomes from playing dice or the roulette. In this book, *risk* refers to a specific type of uncertainty based on an educated understanding of probabilities. But when it comes to long-term uncertainty, such as in the case of climate change, this definition becomes trickier, or even forced. Accepting discussions on the derivation of these probabilities may very well invoke scepticism. Because of this, it is useful to think beyond the traditional understanding of risk (Hansen 2022).

In macroeconomic policymaking, decision-making often involves working under uncertainty. Sometimes it is a matter of not having timely data at hand, while in other instances it may be because of limitations in considering all relevant considerations. Monetary authorities often face the question of adjusting their forecasts or plans in the face of uncertainty. COVID-19 has also shown the challenges of making decisions in the absence of adequate data and the importance of making mid-course adjustments as more complete information becomes available. Policymakers constantly face the trade-off between waiting for measurable information and making timely policy decisions.

Measuring Risk

Risk is a forward-looking idea that captures the eventuality of an occurrence. The concept of risk has been described in several ways, but germane to the definitions is the idea that it is the combination of the probability of an event occurring and its potential consequences or impacts. These consequences can be positive or negative, but in the context of disasters, risks have a primarily negative facet. Disaster risk signifies the chances of adverse effects from the interaction of physical hazards, as well as the exposure and vulnerability to them (next section). In the era of climate change, the danger from hazards can still be understood, in the first instance, as the combined result of the probability of their occurrence on the one side and the damage they cause on the other. But the needed anticipation and the degree of response will be greater for this high risk.

Disaster risk management can be thought of as comprising two related, but distinct areas: (i) disaster risk reduction and (ii) disaster management (IPCC 2012). Disaster risk reduction denotes both a policy objective, and the measures to anticipate and reduce the (intensity of the) hazard, (people's) exposure, and (their) vulnerability. Disaster management refers to the processes involved in designing, implementing, and evaluating approaches to promote and improve disaster preparedness, response, and recovery.

Probability and Impact

The well-known graphic in Fig. 3.2 (analogous to the "Gartner matrix") depicts how the probability of an occurrence rises along one axis, while its impact increases along the other, making for combinations of the two and enabling stakeholders to ascertain a range of low, medium and high risk situations. For example, low probability combined with low impact denotes low-risk events, such as minor allergies that can be taken care of with over-the-counter medication. A high-probability event with low impact can be thought of as a medium risk, for example, contracting the common flu. And a low-probability, and hence unexpected, event, such as a once in a half-century pandemic, despite its enormous impact or damage, can be seen as another type of medium risk. They are sometimes labelled black swans (Taleb 2007).

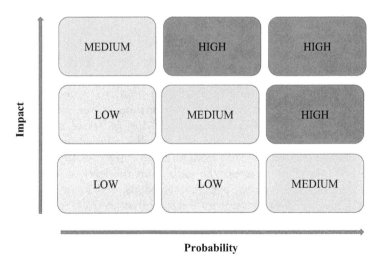

Fig. 3.2 Profiling risk (*Note* The arrows show the impact and the probability going from low to medium to high, and the cells represent low, medium or high risk. *Source* Based on well-known depictions)

However, if the high impact or even the medium impact events become more frequent, they assume the characteristics of high risk. The implications for resilience building change accordingly. They then can be considered high-risk events, which are sometimes named gray rhinos (DeAngelis 2018; Wucker 2016). This formulation of risk can capture the degree of risk in terms of its probability of occurrence and level of impact. Within that framework, a low-probability but high-impact event might be thought of as an intensive risk, while a high-probability but low-impact event would be an extensive risk.

In the wake of climate change, it is the low probability and high impact versus high probability and high impact that catches the most attention. Its implications become clear when thinking of a two-part response, adaptation (like adjustments on the coastlines), and mitigation (like cutting carbon emissions). At first blush, climate change is a continuing crisis compared with the episodic nature of a pandemic. The direct pandemic hit makes the rationale for a quick adaptation clear. But though longer lasting, climate change too presents a series of episodes calling for a pandemic-like quick adaptation at every turn.

But the even bigger concern is mitigation because adaptation without prevention makes it hard to cope with the destruction inherent in ever-increasing calamities (see Chapter 8 also). Despite warnings from scientists and health experts, prevention of future pandemics has not been a top priority compared with dealing with the current crisis. Policy incentives are stacked in favour of dealing with a current problem, in view of its immediate visibility and reward, rather than trying to prevent or ameliorate future ones whose benefits accrue with time. In the case of climate change, mitigation touches on the underlying causes of the problem, which are best confronted early on and not allowed to accumulate.

For climate change, the shift from rare and dangerous to frequent and dangerous is happening across the globe, but the level of impact varies across geographical regions, population groups, and income segments. For example, the destruction from weather disasters is on the rise everywhere, but especially in low-lying coastal regions like the Pacific Islands. Furthermore, the damage is most severe for low-income groups who are both more exposed and more vulnerable to these hazards. Low-income countries are especially pressed to improve their economic performance, but they too are challenged not to pursue carbon-intensive growth that will be counterproductive.

To elaborate on this relationship, the risk of a scenario (RS) can be thought of as a function of the probability of a disaster (PD) and the likely consequence or impact (IM). In a policy context, PD can be qualified by the probability that the event might be averted (PA) (Biringer and Danneels 2001). If the latter is 1, of course, the risk is negated. That is,

$$RS = [PD \times (1 - PA)] \times IM.$$

Disaster risk, therefore, can be depicted in terms of the probability of occurrence and the resulting impact, both informed by experience and changing conditions.

Modelling Risk

In this simple framework, the probability of an event as well as the likely impact can be proxied in the first instance and as a starting point by its past frequency. In the case of food insecurity, for example, the frequency of past shortages in a region and their impact might signal future scenarios. But although past records are instructive, emerging events also need to be accounted for in making projections. For hazards of nature, a rapidly worsening climatic scenario needs to inform risk profiles based on past episodes.

These are simple frameworks to think about the incidence and magnitude of risk. But detailing and modelling the factors that make up risk can be complex because these are applications that differ across sectors and countries and over time. Each component of risk has risk drivers, such as income and poverty, climate change and environmental destruction, and, in the final analysis, society's ability to shape and respond to disasters.

It is one thing to associate contributing factors to certain risks, but it is another to attribute a causal relationship to them. Until the early 2010s, scientists did not attribute any one extreme weather event to climate change, but studies are increasingly indicating that the blame for more extreme temperatures, severe storms, droughts, and floods can be assigned to anthropogenic climate change. On the US West Coast, arson or not putting out a cigarette has been cited as the trigger for forest fires, but it is global warming that makes these fires more frequent and deadly (Rattner and Newburger 2020; Hausfather 2018). In addition, the lack of readiness could have contributed to the destructive force of

Australia's bushfires of 2019 and 2020, but their frequency and ferocity are rising due to climate change (ANU 2020; Gunia 2020). There is growing evidence about the connection between human-caused global warming and extreme disasters as Chapter 6 elaborates.

Sources of Disaster Risks

From a policy perspective, it is important to consider the sources of risk, which can then be influenced. It is widely recognised that disasters can be understood in terms of three main sources of risk: exposure, vulnerability, and intensity (of the hazard) (Fig. 3.3). For example, disaster risks facing people living along low-lying coastlines depend not only on how exposed they are but also on their ability to cope with, say, storms and the extremity of storms. Extreme heat, water scarcity, and rising sea levels may entail the abandonment of several cities and all the sunk capital.

Going to the sources of disaster risks, people's exposure to climate risks is steadily increasing. Their vulnerability to risk, despite important progress in many aspects of resilience, is also on the rise. Most striking, the intensity of climate hazards itself is rising in all regions of the world. The more the intensity of the hazards is predictably on the rise, the greater is the premium on efforts in the pre-disaster phase. But to enable greater investments in preparedness, society needs to value prevention.

Exposure

One significant factor in disaster risk is the exposure of people and assets. Exposure refers to the presence of people, ecosystems, infrastructure, and assets in settings that could be adversely affected by disasters. In the sense a disaster risk is usually understood, exposure is a necessary, but not sufficient, determinant of risk. An intense storm in sparsely populated areas

Fig. 3.3 Contributors to disaster risk (*Source* Kahan 2010; UNISDR 2017; Author's adaptation)

can pose fewer risks than a moderate storm in a densely populated city. If few people and economic resources were in (or exposed to) potentially dangerous settings, the problem of disaster risk, as commonly assessed, would be minimal. But ecological and environmental damages would still occur.

Rising population densities, especially in highly built-up urban centres, have caused more people to move to areas that are danger zones, which in turn aggravates exposure and eventual damages of hazards. The increasing damages from tropical cyclones can be explained by increasing economic activity and population densities in areas prone to cyclones (Mendelsohn et al. 2012). A great deal of policy attention focuses on the best ways to lower exposure, through measures like zoning, evacuation, and locational planning.

A part of exposure is historically determined—for instance, the exposure of fishing communities in coastal areas to storms. It can also be affected by policy, such as zoning regulations modifying the location of households and businesses. Economic considerations, population growth, and urbanisation drive exposure. Hazard-prone areas such as flood plains and coastlines attract economic development leading to a concentration of people and assets over time. Understanding the economic decisions that led more people to live in harm's way is necessary for managing exposure to risks.

Vulnerability

A second aspect is vulnerability, or the propensity to be adversely affected, including sensitivity to harm and a lack of capacity to cope with and adapt to it. It is possible to be exposed but not vulnerable. Vulnerability refers to the capacity of populations to withstand shocks—in other words, how vulnerable they are in terms of education, access to technology, and income levels. Preparedness of a community to recover from a shock depends on, among other things, its past exposure to similar events. Existing income levels may decide in good measure people's financial capabilities and how able they are to withstand a shock.

Policies and investments in infrastructure and human development significantly shape vulnerability. Economic development and higher incomes can help withstand hazards better, but the relationship between income and vulnerability is not linear and is affected by other factors. For instance, environmental degradation that often accompanies rapid income growth accentuates vulnerability. In Southeast Asia, and Asia more broadly, environmental destruction seems to rise initially with income growth before it starts to eventually decline.

Onuma et al. (2017) found that evacuees of the 2016 earthquake in Kumamoto prefecture in Japan—the largest earthquake since the Tohoku earthquake and tsunami in 2011—had difficulties securing survival necessities because of their limited experience with disasters, resulting in lower perceptions of risk and less knowledge of the need to prepare, for example, assembling emergency items. They also found that those with lower income and education were more vulnerable to disaster damage from a lack of preparation, limited ability to ensure preparatory supplies and inadequate understanding of disaster risks.

Vulnerability, like exposure, is also greatly affected by socioeconomic factors. Several studies find that income, education, and institutions influence vulnerability and, subsequently, disaster impacts (Rentschler 2013). A case in point is that flash floods cause far more fatalities in poorer communities than in affluent areas. Poorer segments of the population with meagre resources often end up in higher-risk peripheral areas that can be especially prone to flash floods and often have little protection in poorly built homes. And when livelihoods are affected, losses are further magnified, leaving people even more vulnerable. Hallegatte et al. (2017) envisaged that there would be 26 million fewer people in extreme poverty—that is living on less than US$1.90 a day at the time of the study—if the extremity of disasters could be prevented the following year.

Intensity

In analysing disaster risk, it is also essential to account for the intensity of events. Intensity can be defined by how strong an event or hazard is; for example, storms with windspeeds greater than 100 km/hour. Showing this variable explicitly is valuable for climate change analysis, where new events are systematically more intense than previous ones, making this consideration salient. Often, attention is focused in the wake of a disaster on how prepared the systems were in dealing with it, without recognising that the event in question broke all records in its extremity.

Climate-related intense hazards[1] have unmistakably been on the rise in the past two decades, and Southeast Asia is one of the sub-regions hit the hardest. This could mean an economic reduction of 11% in the region by the end of the century due to damage to livelihoods, and an increased global health risk due to the easier spread of infectious diseases

(Hiebert and Fallin 2021). The socioeconomic fallout from climate disasters is proving to be overwhelming, even though its supremacy does not feature in economic policymaking, like energy prices or inflation.

Changes in the global climatic system due to both natural variability and change that is anthropogenic are affecting the frequency and severity of climate-related hazards, causing extreme events of storms, floods, heatwaves, fires, and droughts. With climate change, the severity of hazards of nature becomes something that can be influenced by policy—that is, the severity of the hazard itself is a subject of policy response.

Disaster Determinants and Policy

Disasters are sometimes thought of as natural phenomena that are beyond human control. But as mentioned, disaster risk is the combined result of the probability of being exposed to a hazard, vulnerability, or ability to withstand it, and the intensity of the hazard itself. The chances of being hit by a disaster clearly have elements of human control, including the degree of people's exposure and their vulnerability. It is the severity of the hazard that is most often taken as a given. But climate change has human fingerprints on it, and therefore disaster policy needs to recognise that the intensity of the disaster itself can be influenced.

Aside from the intensity of a weather event like a tropical storm, the impact of the disaster is heavily influenced by human behaviour too. Soil erosion at beaches around the world contributing to the intensity of hazards is a case in point. The accelerated disappearance of beaches in Greece and Cyprus is attributed more to the construction of new harbours than to climatic factors (Tsoukala et al. 2015). The barrier effect of the port of Valencia, Spain, as in Thiruvananthapuram, India, is partly responsible for beach erosion (Chapapria and Peris 2021; Thomas 2022). Various country studies suggest a strong relation between the growing ferocity of landslides with deforestation, quarrying, and construction.

The human element has huge implications for disaster risk management, especially in its phase of prevention when steps can be taken to avert extreme outcomes. It is vital to appreciate the enormity of disaster risks for policy reforms to respond to the risks. The needs of course vary across sectors. Consider electricity security, for example, that is threatened by weather impacts. Climate change affects power generation, resilience of transmission and distribution networks. Businesses need to be on the frontline in guarding against risks such as power outages, but government policy needs to support building a resilient electricity system (IEA 2021).

Investments that put all the weight on the financial bottom line and excessively discount future risks are not sustainable. Reducing disaster risk cannot also be dealt with in isolation from economic and social considerations. National development plans should establish and build in measures that can identify the enormity of emerging threats, together with policies to steer and support people towards building resilience. Disaster response is associated with governance, civil and political rights. Countries with strong financial institutions, openness to trade, and higher quality of government spending were found to be better able to withstand initial disaster shocks (Toya and Skidmore 2007).

Thus, it is necessary that institutional and adaptive capacity is strengthened in urban areas, especially those that are highly susceptible to flooding, storm surges, and tropical cyclones. The perception and reality of risk are shaped by resilience profiles of locations and the population. The policy questions about resilience building allow decision makers to manage better and lower the country's risk profile. The capability to face, adapt, and respond to risk has features of resilience, and these are taken up in the next chapter.

Conclusions

The overarching conclusion is the need to appreciate the enormity of risks which in turn should shape attitudes to policy reforms. Steps to reduce disaster risk need to become integral to how economic and social considerations are shaped, rather than an afterthought. Aside from having in place preparatory infrastructure and human capacity to deal with the growing challenges, such foresight also allows for the timely disbursement of funds for disaster management as they would already have been planned for. National development plans should establish measures to identify high risks, together with policies to steer and support resilience building.

In this context, it pays to understand the sources of risks in terms of people's exposure, their vulnerability, as well as the intensity of hazards. The onset of climate change is rapidly changing the way categories of risk are to be thought about. High-impact events need to be considered seriously to make educated calculations of whether they are also threatening to become more frequent. In some instances, this calculus needs to be applied at the regional or global levels, as for pandemics or climate change, and in others, they are more country-specific, for example, social cohesion or food security.

In underpinning attitudes to risk, the roles of analysis, perception of the future, and communication come through, as further considered in Chapter 6. Scientists have been careful not to attribute individual natural disasters to climate change, but new studies are indicating that global warming driven by human activities is causing more extreme disasters. If the severity of the hazard itself is endogenous, it becomes a policy variable as well. In making policy, how the future is viewed matters, especially with respect to how benefits and costs accruing in the future might be discounted. The discount rate for climate projects is a major issue, which is taken up in Chapter 7. Communication of this relationship in understandable terms is the responsibility of the scientists, economists, and the media.

NOTE

1. The United Nations Office for Disaster Risk Reduction (UNDRR) (2017) defines natural hazards as events that cause loss of life, injury, health impacts, property damage, social and economic disruption, or environmental degradation. Typhoons, landslides, and earthquakes are rapid-onset hazards. Hazards can also be slow in onset, for instance, sea level rise, coastal erosion, drought, famine, environmental degradation, desertification, and deforestation.

BIBLIOGRAPHY

Australian National University. 2020. "Study Shows Wildfires Increasing in Size and Frequency." https://www.anu.edu.au/news/all-news/study-shows-wildfires-increasing-in-size-and-frequency.

Barnett, Michael, William Brock, and Lars Peter Hansen. 2022. "Climate Change Uncertainty Spillover in the Macroeconomy." In *NBER Macroeconomics Annual*, 253–320. Chicago, IL: University of Chicago Press. https://doi.org/10.1086/718668.

Biringer, Betty, and Jeffrey J. Danneels. 2001. "Risk Assessment Methodology for Protecting Our Critical Physical Infrastructures." In *Risk-Based Decision making in Water Resources IX*, 33–43. Santa Barbara: American Society of Civil Engineers. https://doi.org/10.1061/40577(306)4.

Bremmer, Ian. 2022. *The Power of Crisis: How Three Threats—and Our Response—Will Change the World*. Simon & Schuster.
Burt, Tom. 2021. "Protecting People from Recent Cyberattacks." Microsoft On the Issues. December 6, 2021. https://blogs.microsoft.com/on-the-issues/2021/12/06/cyberattacks-nickel-dcu-china/.
Business Wire. 2021. "IronNet Expands Collective Defense in Singapore to Defend Against Cyberattacks." BusinessWire: A Berkshire Hathaway Company. https://www.businesswire.com/news/home/20210811005380/en/IronNet-Expands-Collective-Defense-in-Singapore-to-Defend-Against-Cyberattacks.
Chapapría, V.E., and J.S. Peris. 2021. "Vulnerability of Coastal Areas Due to Infrastructure: The Case of Valencia Port (Spain)." *Land* 10: 1344. https://doi.org/10.3390/land10121344.
Chivian, Eric, and Aaron Bernstein, eds. 2008. *Sustaining Life: How Human Health Depends on Biodiversity*. Center for Health and the Global Environmen, Harvrd Medical School: Oxford University Press.
Dasgupta, Partha. 2021. *Symposium on The Economics of Biodiversity: The Dasgupta Review*. London: HM Treasury. Afterword. Environmental and Resource Economics.
DeAngelis, Stephen. 2018. "Supply Chain Risk in the Age of Big Data - Enterra Solutions." https://enterrasolutions.com/supply-chain-risk-in-the-age-of-big-data/.
Gunia, Amy. 2020. "Australian Bushfires Will Get Worse With Climate Change: Report|Time." https://time.com/5904762/australia-bushfires-climate-change-report/.
Hallegatte, Stephane, Adrien Vogt-Schilb, Mook Bangalore, and Julie Rozenberg. 2017. *Unbreakable: Building the Resilience of the Poor in the Face of Natural Disasters*. Washington, DC: World Bank. https://doi.org/10.1596/978-1-4648-1003-9.
Hansen, Lars Peter. 2022. "Confronting Uncertainty in Climate Policy." The University of Chicago Booth School of Business. https://www.chicagobooth.edu/review/confronting-uncertainty-climate-policy.
Hausfather, Zeke. 2018. "Factcheck: How Global Warming Has Increased US Wildfires." Carbon Brief. https://www.carbonbrief.org/factcheck-how-global-warming-has-increased-us-wildfires.
Hiebert, Murray, and Danielle Fallin. 2021. "Security Challenges of Climate Change in Southeast Asia." CSIS. https://www.csis.org/analysis/security-challenges-climate-change-southeast-asia.
International Energy Agency. 2021. Climate Resilience. https://www.iea.org/reports/climate-resilience.

Intergovernmental Panel on Climate Change. 2012. *Managing the Risks of Extreme Events and Disasters to Advance Climate Change Adaptation*, 582. Cambridge: Cambridge University Press.

Kahan, Jerome H. 2010. "Risk and Resilience: Exploring the Relationship." Homeland Security Studies & Analysis Institute, November, 156.

Kan, Hai-Dong, Bing-Hen Chen, Chao-Wei Fu, Shun-Zhang Yu, Li-Na Mu. 2005. "Relationship Between Ambient Air Pollution and Daily Mortality of SARS in Beijing". [J]. *Biomedical and Environmental Sciences* 18(1): 1–4.

McKinsey & Company. 2020. "Cybersecurity in a Digital Era." Digital McKinsey and Global Risk Practice. McKinsey & Company. https://www.mckinsey.com/~/media/mckinsey/business%20functions/risk/our%20insights/cybersecurity%20in%20a%20digital%20era/cybersecurity%20in%20a%20digital%20era.pdf.

Mendelsohn, Robert, Kerry Emanuel, Shun Chonabayashi, and Laura Bakkensen. 2012. The Impact of Climate Change on Global Tropical Cyclone Damage. *Nature Climate Change* 2 (3): 205–9. https://doi.org/10.1038/nclimate1357.

Onuma, Hiroki, Kong Joo Shin, and Shunsuke Managi. 2017. Household Preparedness for Natural Disasters: Impact of Disaster Experience and Implications for Future Disaster Risks in Japan. *International Journal of Disaster Risk Reduction* 21: 148–58. https://doi.org/10.1016/j.ijdrr.2016.11.004.

Ranger, Nicola, Olivier Mahul, and Irene Monasterolo. 2021. "Managing the Financial Risks of Climate Change and Pandemics: What We Know (and Don't Know)." *One Earth (Cambridge, Mass.)* 4 (10): 1375–85. https://doi.org/10.1016/j.oneear.2021.09.017.

Rattner, Nate, and Emma Newburger. 2020. "These Charts Show How Wildfires Are Getting Larger, More Severe in the U.S." CNBC. 2020. https://www.cnbc.com/2020/09/18/fires-in-california-oregon-and-washington-data-shows-blazes-getting-worse-.html.

Rentschler, Jun E. 2013. *Why Resilience Matters: The Poverty Impacts of Disasters*. Washington, DC: World Bank. https://doi.org/10.1596/1813-9450-6699.

Ritchie, Rae. 2019. "Maersk: Springing Back from a Catastrophic Cyber-Attack | I-CIO." Global Intelligence for Digital Leaders. https://www.i-cio.com/management/insight/item/maersk-springing-back-from-a-catastrophic-cyber-attack.

Sillmann, J., Christensen, I., Hochrainer-Stigler, S., Huang-Lachmann, J., Juhola, S., Kornhuber, K., Mahecha, M., Mechler, R., Reichstein, M., Ruane, A.C., Schweizer, P.-J. and Williams, S. 2022. *ISC-UNDRR-RISK KAN Briefing Note on Systemic Risk*. Paris: International Science Council. https://doi.org/10.24948/2022.01

Simpson, Nicholas P., and Christopher H. Trisos. 2021. "Guest Post: How to Assess the Multiple Interacting Risks of Climate Change." *Carbon Brief* (blog). April 27, 2021. https://www.carbonbrief.org/guest-post-how-to-ass ess-the-multiple-interacting-risks-of-climate-change/.

Strader, Stephen. 2022. "Opinion: The Hurricane Problem Florida Could Have Avoided." *CNN*, September 30, 2022. https://lite.cnn.com/en/article/h_2 a8b4fd8673ace535abf22f16f2a877f.

Taleb, N. N. 2007. *The Black Swan: The Impact of the Highly Improbable*. Random House. April 17.

Thomas, Vinod. 2022. "For Vizhinjam, Business as Usual Is Not an Option." *The Hindu*, September 7, 2022, sec. Comment. https://www.thehindu. com/opinion/op-ed/for-vizhinjam-business-as-usual-is-not-an-option/articl e65858263.ece.

Toya, Hideki, and Mark Skidmore. 2007. Economic Development and the Impacts of Natural Disasters. *Economics Letters* 94 (1): 20–5. https://doi. org/10.1016/j.econlet.2006.06.020.

Tsoukala, V.K., V. Katsardi, K. Hadjibiros, et al. 2015. "Beach Erosion and Consequential Impacts Due to the Presence of Harbours in Sandy Beaches in Greece and Cyprus." *Environmental Process* 2 (Suppl 1): 55–71. https:// doi.org/10.1007/s40710-015-0096-0.

United Nations International Strategy for Disaster Reduction (UNISDR). 2017. "2017 Global Platform for Disaster Risk Reduction." Mexico. https://www. undrr.org/publication/2017-global-platform-disaster-risk-reduction-procee dings.

United Nations Office for Disaster Risk Reduction. 2022. "Global Assessment Report on Disaster Risk Reduction." https://www.undrr.org/publication/ global-assessment-report-disaster-risk-reduction-2022

World Economic Forum. 2021. "The Global Risks Report 2021." https://www. weforum.org/reports/the-global-risks-report-2021/.

Wu, X., R. C. Nethery, M. B. Sabath, D. Braun, and F. Dominici. 2020. "Air Pollution and COVID-19 Mortality in the United States: Strengths and Limitations of an Ecological Regression Analysis." *Science Advances* 6 (45): eabd4049. https://doi.org/10.1126/sciadv.abd4049.

Wucker, Michele. 2016. *The Gray Rhino: How to Recognize and Act on the Obvious Dangers we Ignore*. St. Martin's Press. April.

CHAPTER 4

Resilience That Shapes Risk

Our Greatest Glory is not in Never Falling, but in Rising Every Time We Fall. Confucius

Risk reduction—for example, setting up better early warning systems or building better drainage—involves reducing exposure to adverse events or vulnerability to the impacts of these events, as well as lessening their intensity. It means being better prepared for hazards, and is part of acquiring the resilience to face hazards. Resilience also goes beyond preparedness and includes the capabilities to adapt to hazards and handle future challenges. Building resilience is about strengthening all these capabilities.

While risk is a measure of danger faced, resilience is the capacity to confront and manage risk. Risk is surrounded by considerable uncertainty and its precise measurement is fraught with difficulties. Resilience is about the functioning of a system with respect to space and time, how it can withstand shocks and recover, and how it can be reconfigured for better system performance (Galaitsi et al. 2022). This chapter discusses how resilience qualifies risk, an important factor for policymaking.

Development organisations perceive resilience as the idea of individuals, communities, and states recovering from hazards, shocks, or stresses without compromising prospects for development (World Bank 2014). The United Nations International Strategy for Disaster Reduction (2012) stresses the need to be able to resist, absorb, accommodate, and recover. The Organisation for Economic Co-operation and Development

(OECD) defines resilience as the "ability of households, communities, and nations to absorb and recover from shocks, while positively adapting and transforming their structures and means for living in the face of long-term stresses, change and uncertainty" (OECD 2022). The OECD's definition seems especially attuned to the climate crisis.

Nature and Phases of Resilience

Resilience can be related to the functioning of systems. It denotes the ability of a system to withstand, cope, and recover from shocks. Resilience is also the capacity to cope with continual setbacks. Building resilience is a dynamic process shaped by preventive or protective capabilities that need to be created and strengthened, which is shaped by investing in them. In varying degrees, this involves ensuring the stability of systems in the face of risks, building the flexibility to adjust, and enabling change for evolving situations.

Post-pandemic thinking invariably features the centrality of resilience in societal responses to daunting problems. Resilience denotes the speed of recovery but that does not necessarily come from sheer robustness to big shocks. The COVID-19 pandemic taught the world that hardiness and risk aversion are not enough, especially as the virus evolves rapidly. Large shocks do not necessarily need to be confronted with pure strength but call for adaptive, flexible, and innovative interventions. Brunnermeier (2021) likens resilience to the ability of a reed to bend and move with the wind, without breaking. In contrast, an oak tree is robust and withstands terrible storms, but eventually breaks. The capacity to adjust and adapt brings home the value of both resiliency and social cohesion.

Equally, in the wake of the pandemic, there is now a greater appreciation of the role of social cohesion in dealing with growing global challenges. This relates to a sense of trust and solidarity to face adversity together. Singapore's and New Zealand's early and decisive responses to the pandemic suggest that the swift implementation of tracing technology, mask and isolation mandates was facilitated by a sense of social responsibility among citizens. There is a premium not only on clear and effective government responses but also civic cooperation rooted in social capital.

Social Capital

It pays to invest in specific capabilities that shape resilience, like preparedness (e.g., evacuation centers); technical competencies, (e.g., meteorological information sharing); and institutional attributes (e.g., coordination among agencies). Investment to build resilience can also have socioeconomic attributes, including the adequacy of resources. Financing is vital in all phases of resilience-building. Time and time again, disaster experiences bring out the importance of societal attitudes and social cohesion reflected in the strength of social capital (Box 4.1).

Box 4.1 Social Capital and Response to the Nepal Earthquake 2015
Social capital is a key element in recovery from and resilience to disasters. The 2015 earthquake of 7.8 magnitude in Nepal was devastating in its impact, leaving more than 6000 dead and 13,000 injured (Plan International UK 2015). Differences in the resilience with which communities responded brings out the role of social capital in terms of bonding and connectivity across people. The experiences, taken together, of three remote Nepali communities affected by the earthquake with varying access to infrastructure, relief, and recovery programmes are a case in point.

Immediately after the earthquake, high levels of bonding and bridging social capital among residents reduced barriers to collective action and helped efforts to rescue and support affected individuals. This social cohesion seemed to diminish, however, once external relief arrived. Marginalised groups with low social capital, like women and the elderly, were less able to access relief items and funding for rebuilding compared with those of higher social status or with political links. This outcome was the result of inadequate pre-disaster programmes to build resilience in disadvantaged or marginalised social groups. This imbalance in investment in preparation is seen in many parts of the world.

Preexisting socio-cultural inequalities, including those driven by weak bonding relationships in families, gender inequalities, and the remoteness of villages, further undermined communities' resilience. Thus, despite short-term community aid, the overall risk profile of the community remained unchanged. The message is clear that disaster response and disaster investment, including disaster relief, should target women and the elderly to improve the resilience of marginalised communities to future disasters. The payoffs to doing so are high across countries.
Source Panday et al. (2021).

Resilience Phases

Resilience parameters can be divided in several ways and with a range of differentiations. Taken together, previous analysis and experience suggest that they fall into three broad phases: prevention, response, and learning (Fig. 4.1). Prevention refers to preparedness and includes activities for raising situational awareness and taking steps to build resistance in anticipation of risks. Response kicks in during the disaster phase, involving the deployment of staff and resources and coordination across departments and institutions. It is vital in this stage to act swiftly. At the same time, protocols and systems need to be in place to ensure follow-up, avoid delays, and redundancy. In the learning phase, the value of capacity building and planning comes through, all along with an eye to affordability.

The degree of resilience sought should be driven not only by the nature of the risk but also its frequency and intensity. And the extent and effectiveness of investments in resilience can, in turn, affect the nature of future risks. Building resilience effectively can be expected to reduce future risk (Combaz 2014). Resilience to climate change means developing the capacity to go beyond simply coping with the disruption and to being able to help bring the climatic system back into balance. It means improving the ability to deal with more extreme weather events—using preventive steps that can tackle present and future challenges. It means

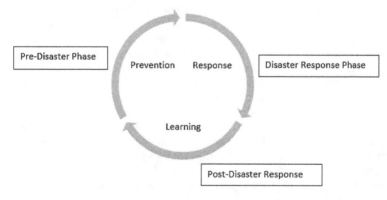

Fig. 4.1 Components of building resilience (*Source* Author's depiction drawing on a vast literature)

modifying the probability of seemingly exogenous shocks and making the risk at least in part endogenous, underscoring the possibility of influencing the impacts through policies and investments.

That distinction applies to different types of intense events that are the focus here, going from acute and immediate ones to chronic and sometimes slow onset ones. Earthquakes and volcanoes, on the one side, and a new virus infection or a debilitating condition, on the other, can be considered exogenous from the point of view of those under attack. Earthquakes are acute and relatively swift in their impact on systems, whereas a pandemic can be far more drawn out if not chronic. In contrast to these exogenous shocks, a run on the banking system, on the one side, and high levels of unemployment, on the other, can be considered endogenous in that they are more the result of ones' actions being taken. The financial crisis could well be acute and swift, whereas unemployment can be drawn out and chronic.

INTERACTION BETWEEN RISK AND RESILIENCE

Policy is most tangible and concrete when responding to an event that has taken place or is taking place, and investing in building resilience to overcome it. More intangible and speculative is investment in anticipation of future perils. The benefits of such investment, as noted before, often do not coincide with election cycles and are in general harder to muster political support for. But the social payoff can nevertheless be large.

Risk-Resilience Scenarios

Figure 4.2 considers four cases of risk-resilience scenarios. In the first case, investments in resilience are going in the wrong direction—resilience decreases and, as a result, the risk also increases. For example, a spike in the use of fossil fuels aggravates future risks of global warming. Or for instance, Levi Strauss, despite committing to a 2025 climate action strategy with a Scope 3 emissions (or indirect emissions by a company's business) target in 2018, found instead that its emissions had increased by 13% from 2016 to 2019 (Eavis and Krauss 2021). This could be due, in part, to the nature of its industry value chain, which relies heavily on coal, diesel, and jet fuel for the manufacture and transport of apparel for sale.

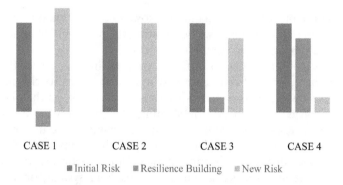

Fig. 4.2 Cases of improving resilience (*Source* Author's depiction)

The second case shows that a lack of investment in resilience, in a situation where little priority is accorded to prevention, leaves the risk profile unchanged. The 2021 study of the 2015 Nepal earthquake in Box 4.1 showed that social capital is a crucial aspect of disaster relief for households. Multiple factors determine the strength of social capital and how sustainable it is. In this case, community resilience was a big contributor to the disaster response, but it began to weaken once external relief became available.

The third case shows the impact of modest investment in resilience building. Risk remains high, possibly as conditions worsen with climate change. A good example is Pakistan's implementation of legal, policy, and institutional regulations to counter future disasters—drawing on lessons learnt, especially from the devastating 2005 earthquake, and seismic risks accruing in the presence of active geological fault lines (Kim et al. 2020). But risks are worsening with climate change obliging countries to go beyond investing in regulatory frameworks and spending on climate proofing infrastructure. The next chapter underscores the unprecedented nature of emerging dangers, exemplified by the 2022 Pakistan floods, for which no amount of traditional risk-resilience calculus would have sufficed.

The final case shows significant investment in resilience tools, such as satellite and radar systems to gauge floods and a corresponding decline in future risk (Box 4.2). Pekalongan city in Java, Indonesia, where risks of floods are high due to two watersheds at its borders, is a good example of this scenario. Because disasters severely damage human livelihoods, the local government and communities are working closely to analyse present data and develop solutions to reduce vulnerabilities to flooding, including through zoning laws, water conservation, and flood-defence infrastructure (Syam et al. 2021).

Box 4.2 gives an example of resilience-building. There is considerable overlap between the idea of resilience and adaptation, but there are some important distinctions too. Resilience is more about building capabilities to anticipate, cope, and recover, while adaptation is more about adjusting and surviving in the face of calamities (GRICCE 2022). With respect to climate change, adaptation means societal adjustments to climate effects. Resilience to climate change, on the other hand, includes the capacity to respond and recuperate from the hazards.

Box 4.2 Resilience-Building in Towns

The development of the Resilience Adaptation Feasibility Tool (RAFT), in response to coastal Virginia's high vulnerability to flooding from rising sea levels in the US, exemplifies resilience building in response to ever-present climate risks. While individual households took precautions to build resilience against flood hazards, a lack of concerted action to shore up governance limited the region's resilience efforts.

The RAFT is a collaborative approach to improving climate resilience using the expertise and resources of different stakeholders to help coastal localities do better in the face of coastal hazards. It is a pre-event activity in the form of a "scorecard" or checklist to advance coastal resilience and was implemented especially to build community resilience in the less-prepared communities in rural coastal areas. The RAFT succeeded in helping Cape Charles, a town in coastal Virginia, create a resilience plan, reducing risks of severe damage and livelihood consequences from future floods.

Another town, Saxis, implemented an action plan to communicate and help its vulnerable households prepare for imminent disasters, thus improving resistance against future disasters. This included building or repairing existing infrastructure to increase disaster resilience. Local leaders also gathered to discuss resilience development based on their own capacities and experiences, relying on their situational awareness, thus building robustness into the action plans within resource and funding means. These were also shared with other localities to promote learning.

Source Andrews et al. (2021).

Resilience Qualifies Risk

Linkov et al. (2013) shows that resilience building goes beyond addressing current risks in the face of emerging and future threats, especially those connected to climate change. Figure 4.3 shows combinations of the extent of risk and the degree of resilience to make the point about the difference the nature of responses makes. Low and high risks are differentiated by the degree of probable departure from the norm, noting that this is hard to pin down. Low and high resilience are marked *inter alia* by the time it takes to get back to the status quo, as country experiences show. The nature of the risk and country conditions relating to the response influence the placing of hazardous events within the four quadrants.

In situations of low or high risk, the ability to recover makes a big difference to the crucial aspects of the functionality of any system, be it a national government or a business enterprise. In that sense, the process of building resilience interacts with the components of risk (exposure, vulnerability, intensity of the hazard) and determines the outcomes in terms of recovery and adaptation (Linkov and Trump 2019). Resilience shapes the degree of risk endured and how effectively a recovery is staged.

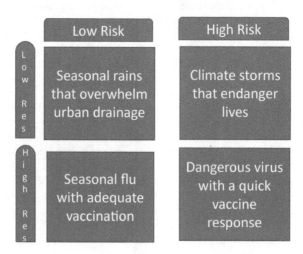

Fig. 4.3 Risk, resilience, and system functionality (*Source* Based on Linkov et al. 2013 and MDBs' project completion reports)

As the next chapter discusses, the key would be going beyond recovering to building stronger capabilities.

It is important to consider the attributes of investing in resilience that allow for quicker and better recoveries. Indonesia and Mexico, whose governments are emphasising resilience building in home improvement subsidy programmes, and Peru, where vulnerability assessments are conducted on properties, are good examples of how governments can build the capabilities of people to limit losses when disasters strike and then recover from these events (The World Bank 2022a).

MAPPING COMPONENTS OF RISK AND RESILIENCE

In understanding the factors that make for stronger resilience, it is useful to view resilience-building in conjunction with risk. It is useful to bring together the main sources of risk discussed earlier, for example, vulnerability, and see how salient dimensions of resiliency affect them. These interactions inform policy responses in disaster management on the part of countries.

Addressing Vulnerability

Three aspects of risk outlined earlier were exposure, vulnerability, and intensity, which now can be juxtaposed with resilience features. Components of resilience have been variously described (see for example Kahan 2010). Resilience attributes in the three phases mentioned earlier, Prevention, Response, and Learning can be elaborated on. Prevention comprises pre-event activities, including building resistance and the ability to absorb shocks. Response can be seen in terms of how resourceful the reaction is and how it deals with restoration. The learning phase includes efforts in developing capacity to deal with future events. All these phases clearly overlap as country experiences show (Fig. 4.4).

Various indices attempt to measure risk on the one hand and resilience on the other, on the basis of which countries are often ranked. It is useful to assess the extent to which a country is at risk from natural hazards. Risk indices capture some or all components of a country's risk, viz, exposure, vulnerability, and intensity (of the hazard)—as well as coping capacity in some instances like a multi-hazard composite index for the Association of Southeast Asian Nations (ASEAN) (The AHA Centre 2018). It is valuable to see the risk profile in relation to policies for augmenting resilience. It

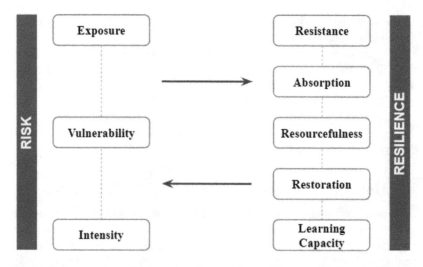

Fig. 4.4 Interaction of risk and resilience (*Source* Based on Linkov and Trump 2019 and the authors' presentations)

would then be valuable to estimate how a country's resilience relative to the hazard changes over time. One such measure, Climate Resilience Policy Indicator, assesses the level of climate resilience of a country by comparing the level of the hazard it faces against its policy preparedness (IEA 2022). According to this measure, considerable differences emerge in terms of the preparedness of nations for climate risks.

The interaction of risk and resilience provides policy insights. Faced with similar hazards, the results of disaster response can vary vastly, as seen in the response to the COVID-19 pandemic across different countries and regions. Even for comparable exposure and vulnerability to shocks, resilience can differ greatly, affecting outcomes. The fragility of the socioeconomic setting, like a country's macroeconomic stability, also has a bearing on how risks are handled. Faced with external shocks, countries are often seen to fortify their own defences.

As the ability to absorb shocks on the resilience side is enabled, vulnerability on the risk side is expected to decline. And as vulnerability is reduced, less resistance to the shocks might be needed. Each of the resilience phases impinges on risk, and vice versa. All these interactions are important. The interplay of vulnerability and resourcefulness is a tangible and identifiable part. The risk presented by floods and storms is partly captured by people's vulnerability that is inherent in the uncertainty of their livelihoods and the fragility of the structures they live in. In turn, actions to lower vulnerability, be it raising incomes or improving safety nets, make for greater effectiveness of the response.

Overcoming vulnerability with elements of resilience building requires prevention and preparedness in all phases of disasters (Sapountzaki 2022). Such an outcome would benefit from attitudes in favour of being prepared for adversities and building contingencies. This approach, however, is more the exception than the norm, partly because of competing priorities. During urgent and immediate concerns, there may be little appetite for allocating capital for resilience strategies whose returns accrue in the future.

In this connection, the idea of antifragility also comes into play (Taleb 2012). Some aspects that shore up resilience gain from the threat of shocks and exposure to adversity. The instinct to find a way out can create fortitude. Some external stress may cause societies to take preventive steps to stave off downsides. As a result, they could bounce forward, signalling greater resiliency. Some societies may not be in favour of taking strong climate steps, but experiences could nudge them into action that is beneficial to themselves and to all. Economies like Singapore or Hong Kong contribute little in terms of total carbon footprint emissions, but have high per capita emissions, and external shocks could prompt them to respond to the bigger picture, in their own interest. Modelling behaviour by the high per capita income economies makes a difference for others' actions. There is also the motivation of each in ensuring that there is a sustainable global future.

The Mekong Delta

Vietnam's agriculture, contributing to 30% of GDP—presents an interesting application (The World Bank 2022b). The Mekong issue, of course, goes beyond Vietnam and affects Laos, Thailand, and Cambodia. Dams impact downstream communities. The Tonle Sap in Cambodia, for instance, is drying up. For Vietnam, sealine incursion of its coasts may be too tough to mitigate. The imperial city of Hue in Thua Thien-Hue province in Vietnam is among the most vulnerable to flooding.

People depending on the Delta's agriculture are mainly rice and fish farmers, many of whom live close to the poverty line. This means they are particularly vulnerable to climate shocks that may affect the productivity of the Delta, given their heavy reliance on its arability. However, the Mekong Delta is also one of the most vulnerable in the world to climate shocks, in terms of coastal erosions and flooding. More than 1.2 million people are highly at risk of threats to livelihoods from climate shocks.

The Mekong Delta Integrated Climate Resilience and Sustainable Livelihoods Project was implemented in 2016 to help farmers and fishers improve their resilience to climatic shocks through tools for climate-smart planning. The project supported increased access to capital, technical assistance to farming practices, and capacity building in their profession for communities and local institutions. Resilience building is the main objective of this project; thus, an all-encompassing hydro-ecological system was implemented, with linked systems that required holistic integrated efforts. Technical assistance was provided by the private sector with relevant expertise. A resilience framework was developed, within which climate-resilient practices were imparted to farmers. Good practices were also gleaned from other global delta initiatives and refined to suit the Vietnamese context.

Country, Regional, and Global Priorities

The priorities for policy responses for resilience building have changed over the years. Some priorities have always been with us, some surface periodically, and some are new concerns. A picture of the changing priorities for resilience emerges from the themes that the World Bank's *World Development Reports* have taken up each year since 1978. Poverty reduction is a perennial and varied theme, including the poverty dimensions of education and finance. The environment was featured in 1992, and climate change in 2010, reflecting the leadership's ambivalent attention to climate change, the biggest development challenge (The World Bank 2022c).

Interlinked with poverty reduction are food security risks, where the poorest countries often face greater threats to accessing sufficient amounts to meet the population's needs. Persistent food insecurity when coupled with other crises (Box 4.3) could turn countries inwards to focus on internal food security, and increasing risks of food shortages, especially for the food-insecure (Deaton and Deaton 2020).

> **Box 4.3 Global Food Insecurity and the Russia–Ukraine War**
>
> Food shortages and food price increases dominate the global socioeconomic outcomes in 2022. Monitoring the developments during the year, Food and Agriculture Organization (FAO) in its various publications has noted "the alarming deterioration of acute food insecurity". International Monetary Fund (IMF) Managing Director Kristalina Georgieva warned that the most serious second-order crisis among the various global crises is that of food insecurity, exacerbated by Russia's war on Ukraine (IMF 2022).
>
> The war has created a ripple effect on food insecurity globally, especially in food-deficient countries like Afghanistan, given that Russia and Ukraine combined contribute to about 30% of global wheat and 20% of corn consumption (Khanna 2022). Restrictions on the export of items like fertilisers and sunflower oil have resulted in global food shortages. This especially hurts the poorest, who spend a large proportion of their incomes on food. The Global Hunger Index warned that over 60 countries would experience significant hunger in 2022, an increase from 47 in the previous year, because of the war (Global Hunger Index 2022).
>
> Globally, the World Food Programme (WFP) assists countries with shortages in food supply. Its support was expected to have risen sharply in 2022 because of the Russia–Ukraine war, highlighting the impact of global shocks on food insecurity, especially for the poorest. Costs aside, social unrest resulting from hunger could wreak havoc on global peace. Coordinated efforts from the WFP, FAO, International Fund for Agricultural Development, and national governments are needed to ensure that policy helps those who need it most.
> *Source* Khanna (2022).

Nearly 350 million people in some 80 countries are estimated to face severe food insecurity in 2022, compared with 280 million in the previous year (WFP and FAO 2022). Food insecurity varies sharply across countries according to a number of factors such as domestic supplies, global shortages and trade, and domestic and trans-boundary conflicts. In the 2022 Global Food Security Index, Finland has the highest ranking among 113 countries with a score of 83.7; Syria has the lowest, at 36.3 (*The Economist* 2022).

Resilience measures to address food security risks are used at national and multilateral levels, with varying degrees of effectiveness. National policies seek to shore up food security and reduce over-reliance on global assistance. Ireland's food status, for example, stems from growing a significant amount of its own food and reducing reliance on imports (Fielding 2020). Ireland also independently collaborates with international organisations on food technologies and processes to improve knowledge-transfer (gov.ie 2021). The World Bank's Food Systems Resilience Program, launched in 2021, for example, supplements national-level programmes aimed at improving food security in West Africa, through technology, information sharing, trade links, and funding (The World Bank 2021).

The sustainability of efforts to build resilience is key to ensuring longer-term viability. Marchese et al. (2018) find that resilience and sustainability are two interlinked components that achieve most benefit when their similarities are capitalised on, and contrasts reduced. A more sustainable system manages to bring about modest dips from recurring shocks. The unsustainable one, while recovering from a deeper dip, returns to yet another deeper dip down the road. When a system's resilience is sustainable, the reduction in its functionality in vital respects is less severe in the face of economic, environmental, and social shocks. This indicates that there is already some resilience in the system to bounce back from shocks.

Natural disasters have an increasingly negative effect on economic growth, and it is instructive to see how the degree of resilience in a country modifies this effect (Tanaka et al. 2021). How a shock, be it a natural hazard or a pandemic, plays out depends importantly on aspects of resilience-building in a country. The COVID-19 experience shows that the impacts of the shock depend on certain pre-shock conditions as well as the nature of the pandemic response. For example, the effect of lockdowns on workers' mobility depended on the pre-pandemic characteristics of a city (Jiang et al. 2022).

In the case of the pandemic, it is also illuminating to observe examples of successes and failures in combating it initially and over time (Box 4.4). The differing experiences bring out the importance of sustaining the efforts and learning from own and others' experiences, noting that the social and financial costs of flip-flops are high. Even with the initial effectiveness of the response, continuing success cannot be taken for granted. There is much to be gained from learning in the face of a dynamic situation.

Box 4.4 Public Health Systems and Resilience
The COVID-19 experience highlights the vital role of public health systems in providing resilience to health risks. Several aspects, such as staff capacity, facilities and supplies, financial resources, and behavioural attributes define the preparedness of health systems. Resilience measures must also be sustainable during a crisis and beyond, as performance of countries are seen to change under changing conditions.

Singapore was ranked by Bloomberg as the most effective among nations in building resilience against COVID-19, with a death rate from infections of 0.05% (Hong 2021). Digital and technological applications responses have been high on Singapore's agenda. Examples are the new generation of apps and technologies like TraceTogether for contact tracing, and a digital check-in system (SafeEntry) to locate the vulnerable and track their travel patterns. Digital technology is also highly relevant for populations at all income levels. For example, Kerala's responses have included the use of E-Sanjeevani, a telemedicine portal, offering psycho-social support for the sick.

New Zealand, ranked second by the Bloomberg index, has made intense use of scientific expertise, spanning public health, infectious diseases, genomics, modelling, and immunology (Geoghegan et al. 2021). Like Singapore, New Zealand has drawn on its lessons from the 2002–2004 Severe Acute Respiratory Syndrome (SARS) outbreak. During the COVID-19 pandemic, a vaccine taskforce was made responsible for ensuring access to safe and effective vaccines as a strategy for exiting the crisis. New Zealand cancelled international flights, while rigorously adhering to public health guidance.

South Korea, facing a massive outbreak of the virus, installed drive-through tests, body sterilisers in public venues and thermal scanners (Aron 2020). Health authorities developed mobile apps to track and monitor those under quarantine and overseas visitors and used drones to disinfect large public areas. South Korea's high teledensity enabled the dissemination of information to target populations through information alerts that were sent to their mobile phones. Noted for its trust in institutions, South Korea has been able to gain from citizen participation as the reach of information and communication technology was combined with the protection of people's privacy.

Conclusions

The interplay of risk and resilience informs policies. The relationship between risk and resilience is dynamic. Stronger resilience means that the risk reflected by an eventuality is lowered. Building resilience to increasing risks like climate change means going beyond just managing risks by taking steps to bring the system back into balance. It also involves building the capacity to be able to confront bigger challenges. Steps that can help with present and future challenges feature strongly in this calculus.

The overarching policy conclusion is about the heightened priority for investing in resilience. These investments feature specific capabilities of staffing, technical competence, and institutional attributes that are key to shaping resilience. Financing is a critical component in all phases of resilience building. And societal attitudes matter in building the required social capital needed to withstand shocks as well. The extent of public support and belief in a country's institutions is a big factor too, as the COVID-19 experience has shown, and the resulting resilience marks a two-way relation with transformative change (Chapter 9).

The juxtaposition of vulnerability to a risk and the cost-effectiveness of resilience- building is a tangible and identifiable part of the risk-resilience interaction. Country efforts can be shaped by identifying the most important components of vulnerability and investing in resilience to withstand them. The sustainability of these efforts becomes all-important because as risks mount, resilience-building strategies need to deepen and extend previous steps. Adaptation adds fortitude in the face of climate disasters, while mitigation lowers the disaster risk, and both are key elements of climate resiliency.

Bibliography

Andrews, Elizabeth, Tanya Denckla Cobb, Michelle Covi, and Angela M. King. 2021. "Communities: Teaming up with Companies, Cities, States, Academia (The RAFT)." In *Collaborating for Climate Resilience*, ed. Ann Goodman and Nilda Mesa, 25–38. London: Routledge. https://doi.org/10.4324/978 0429281242.

Aron, Ravi. 2020. "Combating COVID-19: Lessons from Singapore, South Korea and Taiwan." Knowledge@Wharton. April 21, 2020. https://knowle dge.wharton.upenn.edu/article/singapore-south_korea-taiwan-used-techno logy-combat-COVID-19/.

Brunnermeier, Markus K. 2021. *The Resilient Society*. Princeton Economics.
Combaz, E. 2014. *Disaster Resilience: Topic Guide*. Birmingham, UK: GSDRC, University of Birmingham. https://gsdrc.org/topic-guides/disaster-resilience/concepts/what-is-disaster-resilience/.
Deaton, B. James., and Brady J. Deaton. 2020. "Food Security and Canada's Agricultural System Challenged by COVID-19." *Canadian Journal of Agricultural Economics/revue Canadienne D'agroeconomie* 68 (2): 143–49. https://doi.org/10.1111/cjag.12227.
Eavis, Peter, and Clifford Krauss. 2021. "What's Really Behind Corporate Promises on Climate Change?" *The New York Times*, February 22, sec. Business. https://www.nytimes.com/2021/02/22/business/energy-environment/corporations-climate-change.html.
Fielding, Oscar. 2020. "Ireland's Food Security." ArcGIS StoryMaps. August 28. https://storymaps.arcgis.com/stories/aca3a589eab54b5ba63545957c05ff14.
Galaitsi, S.E., J.M. Keisler, B. Trump, and I. Lincov. 2022. "The Need to Reconcile Concepts that Characterize Systems Facing Threats." *Risk Analysis*. https://doi.org/10.1111/risa.13577.August20.
Geoghegan, Jemma L., Nicole J. Moreland, Graham Le Gros, and James E. Ussher. 2021. "New Zealand's Science-Led Response to the SARS-CoV-2 Pandemic." *Nature Immunology* 22 (3): 262–63. https://doi.org/10.1038/s41590-021-00872-x.
Global Hunger Index. 2022. "Global, Regional, and National Trends." Global Hunger Index (GHI)–Peer-Reviewed Annual Publication Designed to Comprehensively Measure and Track Hunger at the Global, Regional, and Country Levels. 2022. https://www.globalhungerindex.org/trends.html.
gov.ie. 2021. "Minister McConalogue Announces New Agreements on International Knowledge-Sharing by Sustainable Food Systems Ireland." December 7. https://www.gov.ie/en/press-release/c4c4f-minister-mcconalogue-announces-new-agreements-on-international-knowledge-sharing-by-sustainable-food-systems-ireland/.
Governance and Social Development Resource Centre. 2022. "What Is Disaster Resilience?" GSDRC. 2022. https://gsdrc.org/topic-guides/disaster-resilience/concepts/what-is-disaster-resilience/.https://www.chicagobooth.edu/review/confronting-uncertainty-climate-policy.
Grantham Research Institute on Climate Change and the Environment. 2022. "What Is the Difference Between Climate Change Adaptation and Resilience?" September 12. https://www.lse.ac.uk/granthaminstitute/explainers/what-is-the-difference-between-climate-change-adaptation-and-resilience/.

Hong, Jinshan. 2021. "Singapore Is Now the World's Best Place to Be During COVID–Bloomberg." Bloomberg. https://www.bloomberg.com/news/newsletters/2021-04-27/singapore-is-now-the-world-s-best-place-to-be-during-COVID.

International Energy Agency. 2022. "Climate Resilience Policy Indicator." June. https://www.iea.org/reports/climate-resilience-policy-indicator.

International Monetary Fund. 2022. "Statement by IMF Managing Director Kristalina Georgieva on the Publication of the Joint International Financial Institutions Plan to Address Food Insecurity." May 18. https://www.imf.org/en/News/Articles/2022/05/17/pr22158-statement-md-on-joint-plan-to-address-food-insecurity.

Jiang, Yi, Jade R. Laranjo, and Milan Thomas. 2022. "COVID-19 Lockdown Policy and Heterogeneous Responses of Urban Mobility: Evidence from the Philippines." *PLoS One* 17 (6): e0270555. https://doi.org/10.1371/journal.pone.0270555.

Kahan, Jerome H. 2010. "Risk and Resilience: Exploring the Relationship." Homeland Security Studies & Analysis Institute, November, 156.

Khanna, Vikram. 2022. "Russia's War Is Feeding a Hunger Crisis." *The Straits Times*, April 27. https://www.straitstimes.com/opinion/russias-war-is-feeding-a-hunger-crisis.

Kim, Ella, Bilal Khalid, Elif Ayhan, and Ahsan Tehsin. 2020. "Building Seismic Resilience in Pakistan: 15 Years after the 2005 Earthquake." October 8. https://blogs.worldbank.org/endpovertyinsouthasia/building-seismic-resilience-pakistan-15-years-after-2005-earthquake.

Linkov, Igor, and Benjamin D. Trump. 2019. *The Science and Practice of Resilience*. Risk, Systems and Decisions. Cham: Springer International Publishing. https://doi.org/10.1007/978-3-030-04565-4.

Linkov, Igor, Daniel A. Eisenberg, Matthew E. Bates, Derek Chang, Matteo Convertino, Julia H. Allen, Stephen E. Flynn, and Thomas P. Seager. 2013. "Measurable Resilience for Actionable Policy." *Environmental Science & Technology* 47 (18): 10108–10. https://doi.org/10.1021/es403443n.

Marchese, Dayton, Erin Reynolds, Matthew E. Bates, Heather Morgan, Susan Clark, and Igor Linkov. 2018. "Resilience and Sustainability: Similarities and Differences in Environmental Management Applications." *Science of the Total Environment* 613–614 (February): 1275–83. https://doi.org/10.1016/j.scitotenv.2017.09.086.

Organisation for Economic Co-operation and Development. n.d. "Risk and Resilience–OECD." Accessed 4 Mar 2022. https://www.oecd.org/development/conflict-fragility-resilience/risk-resilience/.

Panday, Sarita, Simon Rushton, Jiban Karki, Julie Balen, and Amy Barnes. 2021. "The Role of Social Capital in Disaster Resilience in Remote Communities After the 2015 Nepal Earthquake." *International Journal of Disaster*

Risk Reduction 55 (March): 102112. https://doi.org/10.1016/j.ijdrr.2021.102112.

Plan International UK. 2015. "Nepal Earthquake: Infographic, Facts and Figures." Plan International UK. May 2015. https://plan-uk.org/blogs/nepal-earthquake-infographic.

Sapountzaki, Kalliopi. 2022. Risk Mitigation, Vulnerability Management, and Resilience Under Disasters. *Sustainability* 14 (6): 3589. https://doi.org/10.3390/su14063589.

Syam, Denia, Yoko Okura, and Anna Svensson. 2021. "Integrated Urban Water Resource Management for Climate Resilience: Lessons from Indonesia—Flood Resilience Portal." Flood Resilience Portal. July 2021. https://floodresilience.net/blogs/integrated-water-resource-management-lessons-from-indonesia/.

Taleb, Nassim. 2012. *Antifragile*. Harlow, England: Penguin Books.

Tanaka, Kensuke, Prasiwi Ibrahim, and Ilhame Lagrine. 2021. "Growth Resilience to Large External Shocks in Emerging Asia: Measuring Impact of Natural Disasters and Implications for COVID-19." May 22.

The AHA Centre. 2018. *ASEAN Risk Monitor and Disaster Management Review (ARMOR)*. Jakarta: ASEAN Coordinating Centre for Humanitarian Assistance on Disaster Management (AHA Centre). Available online: ahacentre.org/armor.

The Economist. 2022. "Global Food Security Index (GFSI)." September 2022. https://impact.economist.com/sustainability/project/food-security-index/.

The World Bank. 2014. "Risk and Opportunity, Managing Risk for Development." *World Development Report* 2014: 30.

The World Bank. 2021. "Addressing Food Insecurity and Boosting the Resilience of Food Systems in West Africa." World Bank. November 18. https://www.worldbank.org/en/news/press-release/2021/11/18/addressing-food-insecurity-and-boosting-the-resilience-of-food-systems-in-west-africa.

The World Bank. 2022a. "Global Program for Resilient Housing." World Bank. January 31. https://www.worldbank.org/en/topic/disasterriskmanagement/brief/global-program-for-resilient-housing.

The World Bank. 2022b. "Development Projects : Mekong Delta Integrated Climate Resilience and Sustainable Livelihoods Project." World Bank. August 2022b. https://projects.worldbank.org/en/projects-operations/project-detail/P153544.

The World Bank. 2022c. "WDR Reports." World Bank. 2022c. https://www.worldbank.org/en/publication/wdr/wdr-archive.

United Nations International Strategy for Disaster Reduction. 2012. "UN System Task Team on the Post-2015 UN Development Agenda."

World Food Programme and the Food and Agriculture Organization. 2022. "Hunger Hotspots. FAO-WFP Early Warnings on Acute Food Insecurity:

October 2022 to January 2023 Outlook." Rome. https://www.wfp.org/publications/hunger-hotspots-fao-wfp-early-warnings-acute-food-insecurity-october-2022-january-2023.

CHAPTER 5

New Highs in Risk and Resilience

After climbing a great hill, one only finds that there are many more hills to climb. Nelson Mandela

Whether it is partly from better reporting or from a real spike, disasters, especially climate-related ones, are on the rise, affecting people's anxieties over the global and local outlook. The more such an escalation of risks appears certain, the stronger the case for increasing investments in strengthening resilience. Events are occurring in ways that defy some of the basics of the traditional risk-resilience calculus for wealth and welfare. Sustaining progress as known thus far would seem to go beyond conventional thinking and put to question the frameworks of wealth creation. Following Chapter 2, this chapter looks forward and draws the implications of emerging trends.

Most striking in a world of rising threats is the escalation of global warming affecting every aspect of life on the planet. Levels and trends in the vital climatic variables suggest growing chances of catastrophes (McGuire 2022). The dominant variable as seen in Chapter 2, carbon emissions in the atmosphere, has been rising at over 2 parts per million by volume (ppmv) a year, including the relentless peril from methane. This is despite a fall in emissions over the past decade in several high income countries like Germany and the UK (Statista 2022). Further to Chapter 2 on trends, emissions measured at National Oceanic and Atmospheric Administration's (NOAA) Mauna Loa Atmospheric Baseline

© The Author(s), under exclusive license to Springer Nature
Singapore Pte Ltd. 2023
V. Thomas, *Risk and Resilience in the Era of Climate Change*,
https://doi.org/10.1007/978-981-19-8621-5_5

Observatory peaked in 2022 at 421 ppmv in May, the highest level ever recorded, following a record 419 ppmv a year before.

An annual rate of 2 ppmv a year implies just 15 years to reach the 450-ppmv mark associated with a scenario of warming more than 1.5 °C, and dangerous tipping points. Tipping points denote coordinates where a small further movement produces a vast and permanent effect on critical systems, with deep changes in future trajectories (Gladwell 2000). They are landmark thresholds in climate change, triggering irreversible and negative dynamics, which are much studied by scientists. The current trajectory calls for sizable adaptation investments in national and global investment plans (Roarty and Wheeler 2021). For global temperatures not to exceed 1.5 °C, on the other hand, one order of magnitude is that 7.6% of carbon emissions need to be cut yearly during the decade of 2020s (UNEP 2019; Vaughan 2019)—in contrast to a rising trend.

Rising Risks

Even as the priority given to climate policy lags other concerns, following Chapter 2, worries about the climatic future are uppermost in the minds of businesses, households, and governments. With the epic floods in Southwest US and Pakistan in 2022, rich and poor countries see themselves facing existential threats to their coastlines as well as parts of their inland. Sea levels are rising at 2.5 mm a year on average and 3–4 times as high in places like Miami and Jakarta, as coastal land is also sinking (Nicholls et al. 2021). Coupled with the overuse of groundwater and land subsidence, higher sea levels threaten to submerge coastal areas in many cities. Ho Chi Minh city in the wider Mekong River Delta is a case in point.

Climate migration, already afoot, is likely to strain resources beyond the capabilities of countries (Vince 2022; Von Brackel 2022). There is great asymmetry in the climate profile across regions and income groups. For example, Latin American countries generate less GHG than the US, Europe, Asia, and the Middle East, but are disproportionately hit by droughts, hurricanes, and floods, with a projected 17 million people relocating within 30 years (Phillips 2022). The World Bank (2021) estimated that 216 million people could migrate internally within countries by 2050, a figure that is likely to be revised upwards rapidly. There are similar comparisons for Africa and Latin America on their climate predicament in relation to relatively small carbon footprints.

It could be argued that if the weather damages were directly proportional to carbon emissions by countries, climate action on their part would be swift. But the asymmetry between per person effluents and per person damages across countries is glaring. In the wake of the deadly floods in Pakistan in 2022, the Prime Minister Shehbaz Sharif said at the UN General Assembly: "Nature has unleashed her fury on Pakistan without looking at our carbon footprint, which is next to nothing. Our actions did not contribute to this." The needed decarbonisation calls for investments of sizeable sums (Chapter 8), but raising this financing has been held up, among other things, by this misalignment of cause and effect.

Decades of environmental degradation is exacting a heavy price, including damages from extreme hazards of nature, in all regions of the world. The precipitous loss of biodiversity, described in the next chapter, is aggravated by climate change. The near disappearance of the mighty Aral Sea, a lake lying between Kazakhstan and Uzbekistan and once the world's fourth biggest body of inland water, inter alia due to the diversion and overuse of water for agriculture and the havoc played by global warming is an example. Coral bleaching in Australia's Great Barrier Reef, the biggest coral reef system in the world, due to global warming threatens the survival of corals and marine life. Deforestation in South America's Amazon, the world's largest tropical rainforest, is of acute global concern and has its reasons in forest clearance for livestock and soya cultivation, illegal logging, and government policies of infrastructure. Water crisis and extreme dry conditions caused by overexploitation of land coupled with global warming are leading to the collapse of livelihoods and unprecedented displacement of people in the Horn of Africa. The words of Ellen DeGeneres in the midst of the epic floods in California in early 2023 are prescient: "We need to be nicer to Mother Nature".

Other types of risks are also receiving increasing attention. One area mentioned in Chapter 3 concerns cybersecurity risks related to the advent of technology. The need to protect internet-based systems from cyberthreats is ever-present. Businesses and individuals are investing in guarding against unauthorized access to data. A survey by Allianz Global Corporate and Specialty asked 2650 risk management experts from 89 countries about the biggest business risks for 2022. When it comes to business risks, they ranked cyber incidents the highest. Forty percent of experts surveyed considered cybercrime and business interruption to be the biggest business risks in 2022 (Johnson 2021).

Business entities are also concerned deeply about the carbon transition that they need to make one way or another. Companies are worried about the implied economic risks, for example involving stranded assets. They may react to this by working with governments and international organisations for an orderly transition, and seize new opportunities in the process. But others, for example, coal lobbies, oil and gas majors, also lobby governments to avoid making the transition, which may lessen their own immediate risks but at a high longer-term cost to societies and themselves.

Another example of human-made risk, brought up in Chapter 4, is food insecurity which is a concern increasingly faced by many in the developing world. Nearly a billion people across the world are estimated currently not to get enough food from a health and nutrition viewpoint. The WFP finds that more than 41 million people worldwide are on the brink of famine (Hengel 2021). Africa, Asia, and the Middle East have hotspots of severe undernourishment. With the Russia–Ukraine war, greater food insecurity, stemming from rising food prices and grain and fertiliser shortages, can be expected—and poor countries are the worst hit (Vanderford 2022).

Economic activities periodically face the risk of financial collapse. Given that financial risks are ever-present, it is incumbent on an economy to build and refine its resilience measures after each shock, to ensure that it is better prepared for the next. The Organization of the Petroleum Exporting Countries (OPEC) oil price shock of 1973 was then the biggest financial crisis since the 1929 Great Depression. Subsequently there was the East Asian financial crisis of 1997. The collapse of the US housing market in 2006 led to the global financial crisis of 2007–2008. The default of US subprime loans in 2007 erupted into a crisis that spilled over to the rest of the world. Asia was relatively resilient this time averting a full-blown financial crisis and sharp external adjustments, while Europe was hit hard (Jeasakul et al. 2014).

The experience of the Asian financial crisis in 1997 served to provide lessons for building resilience. Most countries of the region strove to keep external and internal financial balances in check. Credit expansion was modest, reliance on deposit funding was moderate and bank asset quality was improved. Current account deficits and external financing were better managed. Several countries strengthened their stabilisation of macroeconomic accounts. Financial regulations were revamped to oversee and manage risk-taking by households and firms. These and other measures put the financial systems of Asia's economies in good stead to face the next shock to the system, which came around about a decade later. Drawing the right lessons from a crisis can have high rewards.

Shifting Ground

Chapter 2 warned that the rise in global mean temperatures implies disastrous consequences for the sea level. Even under a limit of a 2 °C rise in global temperatures, sea level rise could bring existential threats. A 2022 study found that the loss of ice sheets in Greenland will lead to a minimum of 10 inches (25.4 centimetres) increase in sea levels, and there is little the world can do to stop it, even if emissions were halted immediately, as damage has already been done (Box et al. 2022). Scientists predict this to happen sometime before 2100—and for more extreme weather events to follow, given the accelerating trend of climate change. This calls for major adaptation and changes in layouts and structures for coastal and low-lying centers (Thomas and Quah 2022).

Climate change may seem too overarching in the context of a single episode, but the accumulation of events—from the epic forest fires in the US West Coast to deadly heatwaves in Europe—should make projections of the link tangible (see Chapter 7 also). It is still true that the attribution of a specific disaster to climate change is evolving, and ironclad evidence of causation is not always at hand (CarbonBrief 2021). But the collective evidence from a series of events signals that human-made global warming is the underlying factor behind the increase in the severity and ferocity of extreme events (Harvey 2018). In the triage of responses, climate must be a top priority, along with steps to correct engineering and managerial lapses. Increasingly, policymakers are seeing the fingerprints of human-caused climate change in disasters.

Flash floods in Dallas in August 2022 are a case in point, resulting in deaths, physical infrastructure damage, and communications outages, endangering evacuation, and rescue efforts (Rosenthal et al. 2022). The hurricane season of 2022 in the US East Coast produced unprecedented damage (Philbrick and Wu 2022). In large part, these extreme flash floods were considered to have been caused by global warming due to human activities, which had resulted in a warmer planet and greater atmospheric moisture. Huang and Swain (2022) predict the possibility of a megaflood in California within the next 40 years, at a scale which could result in more than US$1 trillion in economic losses and severely destroy California's lowlands. Like many other states and countries globally, there is a need for California to drastically increase its climate measures, as must many other states in the US and, indeed, countries globally. As noted earlier, extreme winter weather in Europe and the US has also been associated with severe Arctic warming causing "stratospheric polar vortex disruption" (Cohen et al. 2021; also Chapter 6).

Anchoring Resilience

The understanding of resilience, especially with rising climatic risks, calls for much bolder approaches to building and strengthening capacity that acknowledge and act on the root causes of disaster threats. A heightened focus on resilience is crucial to be able to cope with both expected shocks and surprises. The shift in the thinking on resilience is needed to translate into new priorities and more robust budgets. There needs to be transformative change in how resilience is understood, as the final chapter elaborates.

Bigger Dangers, Rebuild Better

With carbon emissions causally linked to these growing disasters, resilience building would have to include far-reaching steps to decarbonise economies. Such measures must become a far bigger component of resilience building, with greater preparedness for more and more predictable catastrophes. This focus on prevention sees the recovery from a calamity as not just being a return to the status quo, but planning for better resilience and emphasising mitigation (Linkov et al. 2013 and related presentations). Figure 5.1 shows the importance of adaptation to improve resilience.

Synergies Across Sectors

A common denominator in disaster readiness is the priority that needs to be given to health and education investments, which are also associated with a good degree of people's trust and citizen participation in institutions. There are several countries like Singapore, South Korea, and New Zealand which have walked an extra mile during COVID-19 through timely investments in health and education, as discussed in Chapter 3. Aside from COVID-19, other infectious diseases, like Ebola, SARS, and the Middle East Respiratory Syndrome (MERS), have in recent years strained the healthcare systems of even developed countries, increasing

5 NEW HIGHS IN RISK AND RESILIENCE   79

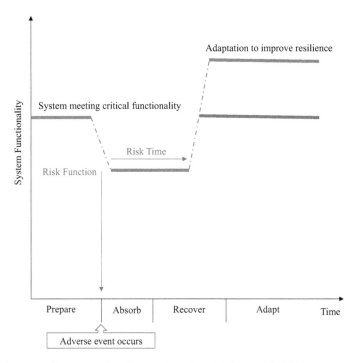

Fig. 5.1 Resilience metrics (*Source* Based on Linkov et al. 2013)

the risks of damage to the affected economies and human development (Nuzzo et al. 2019). Immediate risks include damaging changes to livelihoods and the social fabric. Over time, there could be impacts on health, the environment, geopolitical tensions, and a widening technological divide (Zahidi 2021).

A recent study found that pathogenic human diseases have been exacerbated by global warming, underscoring a deadly link between diseases and the destruction of nature (Mora et al. 2022). It found that 218 out of 375 or 58% of human pathogenic diseases were worsened with the impacts of ten climatic hazards responsive to GHGs. It showed 1006 unique pathways in which climate-related hazards, through different ways of transmission, led to pathogenic diseases. As natural disasters increase in frequency and intensity, humans' ability to cope with the aftermath—especially if resilience is weak—decreases, leaving more room for infectious diseases to spread rapidly. The human pathogenic diseases and transmission pathways exacerbated by climatic risk are too many for societal adaptations to be effective, according to the study. This in turn underscores the urgent need to confront the root cause of the predicament, mitigating carbon emissions, a unifying theme of this book.

The literature on building resilient health systems highlights the need for competent leadership that can coordinate multiple stakeholder responses and provide quality healthcare with capacity to meet growing needs (Kruk et al. 2015). Hong Kong and Singapore present some differences. At the onset of the pandemic in 2020 in Hong Kong, where the only form of resistance to the COVID-19 virus was strict distancing regulations, resilience was not sufficiently built up, resulting in a resurgence of the virus in 2022 (Lew 2022). In Singapore, in tandem with tackling economic and social damage caused by the pandemic, community resilience was built by using technology to engage civilians in various relief initiatives.

This was done through multisector efforts, such as community teaching and formal certification courses for the workforce to improve long-term employability and learning from lessons in social cohesion and digital convenience during the lockdown (IMDA 2020). On a wider scale, the Southeast Asia Health Pandemic Response and Preparedness project, launched by the EU and ASEAN, strengthened healthcare capacity through improved healthcare facilities and access to sanitation, with a focus on marginalised and vulnerable populations (EEAS 2021).

It has long been recognised that climate change places a great strain on land and water and hurts agricultural productivity. There are vast differences in regional and country impact of changing extremes in weather, but the effects, taken together, of global warming on food and agriculture are nothing short of alarming, hurting food supplies and raising food prices (Swinnen et al. 2022; IPCC 2021; Nelson et al. 2009). There are examples of good efforts in weather-resilient agriculture (see Box 4.1), but in view of the emerging food crisis, the priority for finding more widespread solutions for sustainable agriculture is urgent.

Societies differ in their capacity to accurately predict disasters, and many countries, at their national and subnational levels can gain from technologies that strengthen this aspect. But far more than the ability to predict, countries' readiness to cope with setbacks varies widely. With conflicting priorities, countries' attention to investing in climate policies varies. What is clear is that the payoffs from slowing global warming as well as adjusting systems to the inevitable rise in temperatures are high. These investments cut across sectors, from infrastructure to agriculture (see Box 5.1).

Box 5.1 Kenya Climate-Smart Agriculture Project (CSA)
Agriculture is the key income source for about 50% of the Kenyan population, and accounts for almost one-third of Kenya's GDP through meeting the population's food needs and exports. Yet despite the heavy reliance on agriculture, about four-fifths of Kenyan land is highly vulnerable to droughts and is arid or semi-arid.

This means that events like droughts can have a significant impact on livelihoods, due to heavy recovery costs, losses in production assets, and corresponding loss of future income. As climate change intensifies, frequencies of droughts increase, leaving less time for recovery and resilience building in Kenyan communities. The Government of Kenya thus coordinated with the World Bank to implement the CSA project from 2017 to 2022. The goal of the CSA project was resilience building in select smallholder farming and pastoral communities, considering the detriments of climate change. Increased productivity was key. About 522,000 households stood to benefit from this project.

Key aspects of the project were to improve water and soil management, improve access to livestock services, spread knowledge on novel and sustainable agriculture technologies, and improve access to agriculture-related information services to help farmers' decision-making. These strategies were designed to create more sustainable agriculture practices in the long term. There were private stakeholders involved to provide feedback loops and identify evaluation indicators for the project's success. Impact evaluation was provided with a budget of US$4.5 million.

Engaging experts in agriculture practices to share key strategies with farmers, as well as a constant feedback loop from objective impact evaluation, provided for more lasting resilience building. This is due to the on-ground iterative improvement process conducted for the duration of the project. This helped farmers improve the sustainability and resilience of their agriculture practices and mitigate their vulnerabilities to droughts and other effects of climate change, while ensuring that productivity and food security remained unharmed.

Source The World Bank (2017).

Payoffs to Preparedness

Inadequate warning and evacuation were immediate factors in the impact of Typhoon Haiyan in the Philippines in 2013, but high sea levels caused the storm's unprecedented fury (Singer 2014; BBC 2022). In the case of the 2018 floods in the Indian state of Kerala, lapses in weather forecasting and in the management of dams aggravated the flooding, but as during Typhoon Haiyan, the amount of rain was also unprecedented on account of climate change (BBC 2018).

The series of cyclones in Odisha from 1999 to 2019 bring out the value of preventive action, as illustrated in Table 5.1. Deaths due to cyclone damage have reduced significantly over the years due to the massive pre-disaster planning undertaken by the state government. But damages are up sharply over a period of two decades, as in other parts of the world as well. From at least 10,000 deaths in the 1999 cyclone, the fatalities have mostly been in double digits in most of the cyclones including Phailin in 2013 and Fani in 2019. The state authorities took a decision to build upon its capacity, particularly at the community level. Odisha has started community-level warning, built multipurpose cyclone shelters under National Cyclone Risk Mitigation Project, and built an Early Warning Dissemination System with last-mile connectivity that has helped in timely evacuation. The capacity to deal with natural disasters has increased tremendously at the community level.

But in the Philippines, the ever-increasing intensity of superstorms tells a story of past lessons not being sufficient for new highs. While past lessons have helped to prepare for eventualities, the raising of the bar for disasters suggests that more is needed. The past may not be a sufficient guide for the future. Cities that had invested in preparedness have indeed come off better. The Philippines has been at the sharp end of facing strong climate-related dangers. On December 16, 2011,

Table 5.1 Preventive action in Odisha, India (indicative orders of magnitude)

	1999	2013	2019
Cyclone (scale)	Paradip (260 kph)	Phailin (260 kph)	Fani (215 kph)
Deaths	10,000	45	89
Damage	US$4.44 billion	US$4.26 billion	US$8.10 billion

Source Reported numbers based on Government of India estimates; Mohanty (2021)

Typhoon Sendong at 100 kph left 1257 dead in Mindanao and 734 of those deaths took place in the city of Cagayan de Oro. The heavy death toll signified the provinces' unpreparedness as such an exceptional event in December would have occurred once in a thousand years. A year later, on December 4, 2012, Typhoon Pablo at 250 kph—much stronger than the previous one—took 1067 lives in Mindanao. Cagayan de Oro, however, which had invested in early warning and evacuation in the wake of Typhoon Sendong, weathered the storm with hardly any fatality. But none of that could have prepared the country for superstorm Haiyan which hit the province of Samar on November 8, 2013. At 315 kph, this was the most extreme storm ever witnessed. By official count, more than 6000 were killed.

Table 5.2 illustrates the deaths in succeeding years from 2011 to 2013. Steps taken to improve are no match when stronger and deadlier shocks—especially unprecedented ones—happen. Experiences indicate that, especially with increasing frequency and intensity of the events, risk management tools alone are not enough and it is necessary to pay attention to social dimensions that make up a disaster (Tellman and Eakin 2022).

Rising and new dangers bring out the value of innovating even during a crisis. Innovating thinking and practice are needed in specific areas. All countries are strapped for resources, albeit in varying degrees, and so the premium is on innovation and efficiency of the needed interventions. As no country or locality has enough staff and financial resources to face the new highs of risks, there is a need, as learnt during COVID-19, for mobilisation across boundaries, ready to be deployed in times of need.

Given its history and geography, Japan had already built considerable resilience against tsunami and earthquake disasters. The country epitomises "bending adversity" seen, for example, following the earthquake, tsunami, and nuclear meltdown of 2011 (Pilling 2014). This threefold disaster brought out the extreme vulnerability of the country as well as

Table 5.2 Superstorms in the Philippines (rough and indicative estimates)		*2011*	*2012*	*2013*
	Storms	Sendong (100 kph)	Pablo (250 kph)	Haiyan (315 kph)
	Deaths	1257	1067	>6000

Source ABS-CBN News (2014)

the resilience it has built over the years. It has translated into concrete spending on disaster management of amounts that are among the highest as shares of GDP in the world and in the mental makeup and attitudes to resilience.

Sometimes events oblige societies to go beyond the norm, innovating, and even breaking rules as disasters display features that were not anticipated and for which the preparation had fallen short. Japan had gone the extra mile in building community resilience—making sure that citizens knew the importance of preparation during peace times with earthquake lessons from kindergarten onwards (Greenblatt 2011). But the 2011 Tōhoku earthquake and tsunami that hit Japan was unprecedented in its unexpectedness, ferocity, and damage. Previous earthquakes had given Japan plenty of lessons for readiness, but none of them could have given a blueprint to adequately tackle what was to unfold. The disaster left 450,000 homeless and 15,500 dead, on top of severely destroying the built environment (National Geographic 2020).

The experience of junior high school students in Kamaishi City, Iwate Prefecture, brings home the value not only of preparation and drills but also attitude-oriented disaster education, with a focus on children's independence (Katada and Kanai 2016). It shows the value of adapting to previous learning and even breaking established norms of disaster management. The junior high student helped elementary school children to evacuate safely. The norm was "Run to a Hill", but in the face of a record storm, the children took the initiative to seek higher ground. Five lost their lives amidst a high death toll.

It is harder to design interventions for large-scale calamities because of their complexity. But that is what is needed as it remains true that prevention is better than cure. The longer the problem, be it a pandemic or global warming, is allowed to fester, the more complicated the solutions would be. Offsetting this concern would be new technological innovation which can make responses to these risks more effective and less costly.

Confronting Downward Spirals

In some cases that involve wicked problems, society has been able to come together and find ways of working together.[1] Some areas of the tragedy of the commons, for example, overfishing, a common ground of understanding can be reached that would enable the application of norms that enhance the prospects for fishing for all (Boyle 2022). For climate change, convergence to a more sustainable path would be when businesses make the operational connection between disasters and high carbon activities, decarbonise, and help stem global warming.

Divergent tendencies threatening a downward spiral are a great cause for concern, as considered further in Chapter 9. Higher demand for energy or shortages of energy, triggered by weather extremes, could increase the use of readily available coal and other fossil fuels driving up emissions. Since 2000, the world has doubled coal-fired power capacity (Carbon Brief 2022). Efforts to boost standards of living by expanding carbon-intensive energy and infrastructure can aggravate the already deteriorating ecosystems. There could be a heightened worry over energy security. This scenario poses tough choices when low-income countries are also pressed to raise incomes and well-being.

Environmental impact assessments and environmental and social safeguards are means to protect against errors that compound socioenvironmental hazards. Through these assessments, the negative impacts of the project on the environment are meant to be studied and coupled with enforcement of mitigating measures to protect the environment (Transportation Engineering 2017). Environmental protection, elaborate as it can be, needs to be applied seriously across countries. Inadequate safeguards, especially in the face of the climate crisis, make the environmental and economic outlook precarious, as several ill-thought-out development projects show. Many highly visible development projects have been associated with serious socio-environmental damages. Examples include China's Three Gorges Dam completed in 2006, and India's Tata Mudra Thermal Power Generation Project completed in 2013 are examples of socio-environmental harms from infrastructural projects.

Where socio-environmental risks are high, mid-course reviews of a project's progress can help. For example, a US$1 billion project, the Vizhinjam international seaport, authorized in 2015 and now under construction by the Adani Group. With scant environmental and social safeguards, the project is harming the ecology and livelihoods in India's southwestern state of Kerala, my home state that is known for its stunning landscape and exceptional indicators of social progress (Thomas 2022). Global warming is causing sea-levels to rise and coastlines to recede across the world but dredging and sand mining without safeguards are speeding coastal erosion by altering wave patterns and sediment movement. It is the building of the port without even the bare minimum of environmental and social safeguards in one of the most fragile areas in the world that is in good measure responsible for a disappearing coastline and socio-ecological destruction.

In the case of pollution from energy sources, emission inventories can be used to quantify the emission of pollutants associated with a project. Risk assessments can analyse the effects of these pollutants on health. Process hazard analysis involves assessing the potential impacts of *unplanned* hazardous materials. A team may rank the possible hazards and focus on averting those that can cause the most harm. However, these protections are far from adequately done across countries and across sectors. If these unplanned externalities are considered in a holistic assessment of the true impact of such activity, the net economic benefit falls. Efforts must be made to employ methodologies that can paint an accurate picture of the externalities so that remedial measures can be taken.

The so-called big data can be harvested to understand the impacts of negative externalities on households, communities, and the environment. Thomas and Chindarkar (2019) evaluates the usefulness of big data in disaster management, which can provide timely information to help policymakers make quicker decisions to mitigate the effects of a crisis. In the larger context of meeting the United Nations' SDGs, evaluating impacts on social inclusion and environmental governance can align socioeconomic policies more closely with meeting these goals.

Box 5.2 China's Belt and Road

Infrastructure investments usually represent the biggest chunk of spending among economic sectors across countries. They also present the biggest social and environmental damage. Information regarding safeguards that provide social and environmental protection are available in most countries. But in the name of rapid growth, countries from Brazil to India are in a process of sidestepping these much-needed safeguards.

China's Belt and Road Initiative (BRI) is a multi-trillion road investment programme spread across Asia and the globe. Environmental concerns include habitat loss, increased pollution, and the destruction of biodiversity. BRI corridors would cut through important bird sanctuaries or key biodiversity areas, and biodiversity hotspots or global 200 Ecoregions (Audubon 2022; UNWCMC 2020; WWF 2012). These precious resources, because of their inadequate market valuation, are not considered nearly enough in the cost–benefit calculus of the infrastructure projects.

> Road development would create direct and indirect impacts, such as habitat loss, fragmentation, poaching, and logging. In the marine environment, increased sea traffic exacerbates the movement of invasive species and pollution. Climate impacts include the continuation of fossil fuel use and carbon emissions. These risks are costly and do not seem to be adequately or seriously factored into the calculations of the costs and benefits of the investment. On the positive side, the government says that exports of renewable sources of energy that the program enables, in addition to the generation of economic activities, will be a plus.

Conclusions

The readiness of societies to cope with disasters varies widely based on historical precedents and investments being made. The quality of investments in disaster management, including infrastructure, health, and education, matters a great deal. There is also a need for imaginative leadership to coordinate multiple stakeholder responses. The evidence is clear that cities that invested well in disaster preparedness came off vastly better under similar circumstances of recurring adversities.

The emerging new highs in risk oblige policymakers to lift the game in shaping resilience beyond rebuilding. While natural disasters are rooted in nature, their intensity and the resulting damages are also influenced by people's decisions and actions, and the relentless pressure human populations exert on nature. Therefore, innovative investments must become a far bigger component of resilience building, with greater preparedness for increasingly more predictable catastrophes.

Looking ahead, disaster response will need to be dynamic, especially in the absence of automatic and systemic corrections. As noted, global warming hurts energy supply, in turn raising energy prices and putting pressure to use fossil fuels, further aggravating warming. Even in the case of daunting problems such as the tragedy of the commons, forces that make for a convergence to a solution are often found eventually. For climate change, there are forces that lead to divergence. It would be necessary to watch for crosscurrents that aggravate the problem, and there may need to be circuit breakers that stop adverse events spiralling out of control, as suggested in Chapter 9.

Note

1. Rittel and Webber (1973) define wicked problems as problems with no definitive formulation, no criteria to tell the problem-solver when a solution has been found, no true–false or good-bad solutions, no immediate or ultimate test of the solution's efficacy, where every solution is a "one-shot operation" due to lack of ability to trial and error, no exhaustive set of solutions, are unique, are a symptom of another problem, have many attributable explanations for its cause, and that planners cannot afford to get the solution wrong.

Bibliography

Alto Broadcasting System and Chronicle Broadcasting Network News. 2014. "A Tale of Three Disasters, the Wrath of Sendong, Pablo and Yolanda|ABS-CBN News." ABS-CBN News. https://news.abs-cbn.com/specials/3disasters.

Audubon. 2022. "Important Bird Areas." Audubon. https://www.audubon.org/important-bird-areas.

Brackel, Benjamin von. 2022. "Nowhere Left to Go: How Climate Change Is Driving Species to the Ends of the Earth." The Experiment.

British Broadcasting Corporation. 2018. "Why the Kerala Floods Proved So Deadly." *BBC News*, sec. India. https://www.bbc.com/news/world-asia-india-45243868.

British Broadcasting Corporation. 2022. "Tropical Cyclone Case Study–Typhoon Haiyan–Tropical Cyclones–Edexcel–GCSE Geography Revision–Edexcel." BBC Bitesize. https://www.bbc.co.uk/bitesize/guides/z9whg82/revision/4.

Box, Jason E., Alun Hubbard, David B. Bahr, William T. Colgan, Xavier Fettweis, Kenneth D. Mankoff, Adrien Wehrle, et al. 2022. "Greenland Ice Sheet Climate Disequilibrium and Committed Sea-Level Rise." *Nature Climate Change*. https://doi.org/10.22008/FK2/D5JEZ0.

Boyle, Michael J. 2022. "Tragedy of the Commons Definition." Investopedia. https://www.investopedia.com/terms/t/tragedy-of-the-commons.asp.

CarbonBrief. 2021. "Mapped: How Climate Change Affects Extreme Weather around the World." Carbon Brief. https://www.carbonbrief.org/mapped-how-climate-change-affects-extreme-weather-around-the-world.

Carbon Brief. 2022. *Global Coal Power.* https://www.carbonbrief.org/mapped-worlds-coal-power-plants/.

Cohen, Judah et al. 2021. "Linking Arctic variability and change with extreme winter weather in the United States." *Science* 373 (6559). September 1. https://www.science.org/doi/10.1126/science.abi9167.

European External Action Service. 2021. "EU and ASEAN Launch 'Southeast Asia Health Pandemic Response and Preparedness' Project." Text. EEAS–European External Action Service–European Commission, June 2. https://eeas.europa.eu/delegations/association-southeast-asian-nations-asean/99404/eu-and-asean-launch-southeast-asia-health-pandemic-response-and-preparedness%E2%80%9D-project_en.
Gladwell, M. 2000. *The Tipping Point: How Little Things Can Make a Big Difference*. New York: Little Brown.
Greenblatt, Alan. 2011. "Japanese Preparedness Likely Saved Thousands." *NPR*, March 13. https://www.npr.org/2011/03/11/134468071/japanese-preparedness-likely-saved-thousands.
Harvey, Chelsea. 2018. "Scientists Can Now Blame Individual Natural Disasters on Climate Change." Scientific American. https://www.scientificamerican.com/article/scientists-can-now-blame-individual-natural-disasters-on-climate-change/.
Hengel, Livia. 2021. "Famine Alert: How WFP Is Tackling This Other Deadly Pandemic|World Food Programme." https://www.wfp.org/stories/famine-hunger-un-world-food-programme-united-nations.
Huang, Xingying, and Daniel L. Swain. 2022. "Climate Change Is Increasing the Risk of a California Megaflood." *Science Advances* 8 (August). https://doi.org/10.1126/sciadv.abq0995.
Infocomm Media Development Authority. 2020. "Accelerating Nationwide Digitalisation to Build a World-Class Resilient Digital Future." Infocomm Media Development Authority, June 4. http://www.imda.gov.sg/news-and-events/Media-Room/Media-Releases/2020/Accelerating-Nationwide-Digitalisation-To-Build-A-World-Class-Resilient-Digital-Future.
Intergovernmental Panel on Climate Change. 2021. "Climate Change 2021: The Physical Science Basis." Switzerland: Intergovernmental Panel on Climate Change. https://www.ipcc.ch/report/ar6/wg1/downloads/report/IPCC_AR6_WGI_SPM_final.pdf.
Jeasakul, Phakawa, Cheng Hoon Lim, and Erik Lundback. 2014. "Why Was Asia Resilient? Lessons from the Past and for the Future," 48.
Johnson, Joseph. 2021. "Topic: U.S. Companies and Cyber Crime." Statista. http://www.statista.com/topics/1731/smb-and-cyber-crime/.
Katada, Toshitaka, and Masanobu Kanai. 2016 October. "The School Education to Improve the Disaster Response Capacity: A Case of 'Kamaishi Miracle'." *Journal of Disaster Research* 11 (5): 845–56.
Kruk, Margaret E., S. Michael Myers, Tornorlah Varpilah, and Bernice T. Dahn. 2015. "What Is a Resilient Health System? Lessons from Ebola." *The Lancet* 385 (9980): 1910–12. https://doi.org/10.1016/S0140-6736(15)60755-3.

Lew, Linda. 2022. "Singapore vs. Hong Kong: COVID Strategies Push Rivals Further Apart." Bloomberg.Com, February 22. https://www.bloomberg.com/news/articles/2022-02-22/singapore-is-winning-in-hong-kong-s-all-out-fight-against-COVID.

Linkov, Igor, Daniel A. Eisenberg, Matthew E. Bates, Derek Chang, Matteo Convertino, Julia H. Allen, Stephen E. Flynn, and Thomas P. Seager. 2013. "Measurable Resilience for Actionable Policy." *Environmental Science & Technology* 47 (18): 10108–10. https://doi.org/10.1021/es403443n.

McGuire, Bill. 2022. *Hothouse Earth: An Inhabitant's Guide*. Icon Books.

Mohanty, Debi. 2021. "A Super Cyclone Acts as Odisha's Cue to Acing Disaster Management." UNDRR, May 31. https://www.preventionweb.net/news/super-cyclone-acts-odishas-cue-acing-disaster-management.

Mora, Camilo, Tristan McKenzie, Isabella M. Gaw, Jacqueline M. Dean, Hannah von Hammerstein, Tabatha A. Knudson, Renee O. Setter, et al. 2022. "Over Half of Known Human Pathogenic Diseases Can Be Aggravated by Climate Change." *Nature Climate Change*, August 1–7. https://doi.org/10.1038/s41558-022-01426-1.

National Geographic. 2020. "Tohoku Earthquake and Tsunami." National Geographic Society, April 6. http://www.nationalgeographic.org/thisday/mar11/tohoku-earthquake-and-tsunami/.

Nelson, Gerald C., Mark W. Rosegrant, Jawoo Koo, Richard D. Robertson, Timothy Sulser, Tingju Zhu, Claudia Ringler, et al. 2009. "Climate Change: Impact on Agriculture and Costs of Adaptation." *International Food Policy Research Institute (IFPRI)* 2009. https://doi.org/10.2499/0896295354.

Nicholls, Robert J., Daniel Lincke, Jochen Hinkel, Sally Brown, Athanasios T. Vafeidis, Benoit Meyssignac, Susan E. Hanson, Jan-Ludolf. Merkens, and Jiayi Fang. 2021. "A Global Analysis of Subsidence, Relative Sea-Level Change and Coastal Flood Exposure." *Nature Climate Change* 11 (4): 338–42. https://doi.org/10.1038/s41558-021-00993-z.

Nuzzo, Jennifer B., Diane Meyer, Michael Snyder, Sanjana J. Ravi, Ana Lapascu, Jon Souleles, Carolina I. Andrada, and David Bishai. 2019. "What Makes Health Systems Resilient against Infectious Disease Outbreaks and Natural Hazards? Results from a Scoping Review." *BMC Public Health* 19 (1): 1310. https://doi.org/10.1186/s12889-019-7707-z.

Philbrick, Ian Prasad and Ashley Wu. 2022. "Population growth is supercharging the costs of hurricanes." *The New York Times*. The Morning. December 2.

Phillips, Gregory. 2022. "How Climate Change Is Changing Latin America." *Duke Today*, April 26. https://today.duke.edu/2022/04/how-climate-change-changing-latin-america.

Pilling, David. 2014. *Bending Adversity: Japan and the Art of Survival*. Penguin Books. ISBN 10: 0143126954.

Rittel, Horst W. J., and Melvin M. Webber. 1973. "Dilemmas in a General Theory of Planning." *Policy Sciences* 4 (2): 155–69. https://doi.org/10.1007/BF01405730.
Roarty, Dan, and David Wheeler. 2021. "Carbon Handprints: A Climate Positive Framework for Equity Investors." June 22. https://www.bernstein.com/our-insights/insights/2021/articles/carbon-handprints-a-climate-positive-framework-for-equity-investors.html.
Rosenthal, Zach, Mary Beth Gahan, and Annabelle Timsit. 2022. "At Least One Dead after Dallas Area Hit by 1-in-1,000-Year Flood". *Washington Post*, August 22. https://www.washingtonpost.com/nation/2022/08/22/dallas-texas-flash-floods/.
Singer, Meagan. 2014. "2013 State of the Climate: Record-Breaking Super Typhoon Haiyan|NOAA Climate.Gov". https://www.climate.gov/news-features/understanding-climate/2013-state-climate-record-breaking-super-typhoon-haiyan.
Statista. 2022. *Energy and Environment*. https://www.statista.com/statistics/270499/co2-emissions-in-selected-countries/.
Swinnen, Johan, Channing Arndt, and Rob Vos. 2022. "IFPRI Global Food Policy Report 2022: Accelerating Food Systems Transformation to Combat Climate Change|IFPRI : International Food Policy Research Institute." May 12. https://www.ifpri.org/blog/ifpri-global-food-policy-report-2022-accelerating-food-systems-transformation-combat-climate.
Tellman, Beth, and Hallie Eakin. 2022, August 3. "Risk Management Alone Fails to Limit the Impact of Extreme Climate Events." *Nature* 608 (7921): 41–43.
The World Bank. 2017. "World Bank Resilience M&E (ReM&E): Good Practice Case Studies." World Bank Group. https://openknowledge.worldbank.org/bitstream/handle/10986/28387/119939-WP-PUBLIC-P155632-28p-ReMECasestudiesFinal.pdf?sequence=1&isAllowed=y.
The World Bank. 2021. "Climate Change Could Force 216 Million People to Migrate Within Their Own Countries by 2050." World Bank, September 13. https://www.worldbank.org/en/news/press-release/2021/09/13/climate-change-could-force-216-million-people-to-migrate-within-their-own-countries-by-2050.
Thomas, Vinod. 2022. "Adani's Vizhinjam Port Needs Midcourse Change as Damage Grows." Devex, October 10. https://www.devex.com/news/opinion-adani-s-vizhinjam-port-needs-midcourse-change-as-damage-grows-104136.
Thomas, Vinod, and Euston Quah. 2022. "Singapore Needs to Front-Load Its Climate Spending." *The Straits Times*, May 18. https://www.straitstimes.com/opinion/singapore-needs-to-front-load-its-climate-spending.

Thomas, Vinod, and Namrata Chindarkar. 2019. *Economic Evaluation of Sustainable Development*. Singapore: Springer Singapore. https://doi.org/10.1007/978-981-13-6389-4.
Transportation Engineering. 2017. "Mitigating Measure–An Overview|ScienceDirect Topics." https://www-sciencedirect-com.libproxy1.nus.edu.sg/topics/engineering/mitigating-measure.
United Nations Environment Programme. 2019. "Cut Global Emissions by 7.6 Percent Every Year for Next Decade to Meet 1.5 °C Paris Target–UN Report." UN Environment, November 26. http://www.unep.org/news-and-stories/press-release/cut-global-emissions-76-percent-every-year-next-decade-meet-15degc.
United Nations Environment Programme World Conservation Monitoring Centre. 2020. "Key Biodiversity Areas (KBA) Definition| Biodiversity A-Z." https://www.biodiversitya-z.org/content/key-biodiversity-areas-kba.
Vanderford, Richard. 2022. "Rising Food Prices Could Become a Business Risk, Analysts Say." *Risk and Compliance Journal, Wall Street Journal*, August 11. https://www.wsj.com/articles/businesses-should-prepare-for-risk-of-civil-unrest-from-food-scarcity-11660210202.
Vaughan, Adam. 2019. "UN Report Reveals How Hard It Will Be to Meet Climate Change Targets|New Scientist." *NewScientist*. https://www.newscientist.com/article/2224539-un-report-reveals-how-hard-it-will-be-to-meet-climate-change-targets/.
Vince, Gaia. 2022. *Nomad Century: How Climate Migration Will Reshape Our World*. Flatiron Books.
World Wildlife Fund. 2012. "Global 200|Publications|WWF." World Wildlife Fund. https://www.worldwildlife.org/publications/global-200.
Zahidi, Saadia. 2021. "After COVID-19 We Need to Build More Resilient Countries." *Time*, January 19. https://time.com/5930654/COVID-19-after-countries-build-resilient/.

PART II

The Climate Catastrophe

CHAPTER 6

Intractability of Climate Change

Our planet is burning.... We need to hold fossil fuel companies and their enablers to account. António Guterres

When it comes to risk and resilience, climate change presents a quandary of the toughest kind. More is learnt of its growing tenacity and damage over time while solutions to the problem remain elusive. The gap continues to rise between scientific prescriptions of what needs to be done and what is being done to rein in climate change. There already are indications that "hard limits to adaptation have been reached in some ecosystems (high confidence)" (IPCC 2022a).

To avoid mounting loss of life, biodiversity, and infrastructure, ambitious, accelerated measures are required to adapt to climate change, at the same time making rapid, deep cuts in GHG emissions. So far, progress on adaptation is woefully slow as other priorities crowd countries' spending agendas. And there are glaring and growing gaps in steps to mitigate climate change, where investments take on a global character. Building on the evolution of the problem outlined in Part 1, this chapter looks for openings that primise a way forward.

© The Author(s), under exclusive license to Springer Nature Singapore Pte Ltd. 2023
V. Thomas, *Risk and Resilience in the Era of Climate Change*,
https://doi.org/10.1007/978-981-19-8621-5_6

Unheeded Warnings

The continuous divergence between the posture of the generators of global warming and the needs of science based solutions, signals the elusiveness of the climate problem. Other daunting challenges too are complex, but they have shown degrees of convergence in finding solutions. As carbon emissions increase relentlessly, the question is if lessons can be learnt in time and applied fast enough to turn things around. Of concern is that even as the evidence for climate debacles mount, climate measures to avoid catastrophes are triaged below today's task of cleaning up.

Global warming was first researched more than 100 years ago. It was in the nineteenth century when scientists first showed that CO_2 would heat up the planet. Svante Arrhenius' 1884 finding led to the understanding that if you doubled the amount of CO_2 in the atmosphere, it could raise the world's temperature by 5 to 6 degrees Celsius (Thompson 2019). In the last few decades, there have been early warnings of what is to come. In 1988, James Hansen—then a NASA scientist—famously told the US Congress, "The evidence is pretty strong that the greenhouse effect is here" (Shabecoff 1988).

Decades of Red Alerts

In 1993 the World Meteorological Organization (WMO) issued the first state of the climate report (WMO 1994). IPCC reports regularly flag the crisis to get the attention of decision makers and the public. Climate change continues to get worse despite the diagnosis of the problem for decades (Fountain 2019; IPCC 2022b). Wagner and Weitzman (2015), among others, argued that the slower action is taken, the higher the likelihood that extreme weather events will get more frequent and more intense. Indeed, that prediction is being borne out. The three biggest GHGs, carbon dioxide, methane, and nitrous oxide, all registered new highs in 2021, with damaging implications for the trajectory of climate change and disasters (WMO 2022).

The gaps between knowledge of the deleterious effects on the one side and emission trends on the other are extraordinary (Ritchie et al. 2020). At the start of the 2020s, over 35 billion tonnes of carbon dioxide (CO_2) a year are being emitted, together with other hazardous discharges such as methane and nitrous oxide (NO_2). Studies also show that the result is a steady rise in temperatures. Databanks at the local, regional, and global levels paint an analogous and incontrovertible picture. Various data sources depict the trajectory of global warming that is threatening economic progress and civilisation itself. Rising temperatures is humanity's existential crisis.

Concurrently, it is also clear that the mean sea level has risen about 8–9 inches (20–23 centimetres) since 1880, with about a third of that attributable to just the last two and a half decades. The rising water level is mostly due to a combination of meltwater from glaciers and ice sheets and the thermal expansion of seawater as it warms. In 2020, the global mean sea level was 3.6 inches (91.4 millimetres) above the 1993 average, making it the highest annual average in the satellite record (1993-present). By the end of the century, the global mean sea level is likely to rise at least one foot (0.3 metres) above 2000 levels, even if GHG emissions follow a relatively lower trend in the coming decades.

Reflecting the elusiveness of the problem, extreme and more frequent weather events are causing cascading impacts across all regions of the world. The global mean temperature is rising, and the rise in average sea levels is alarming. A record number of floods, hurricanes, and wildfires intensified by climate change cost an estimated US$210 billion in damages globally in 2020 (Newburger 2021). The intractability of the climate conundrum is reflected in the continuation of these dangerous trends despite all the scientific evidence. New projections suggest that the worst is yet to come.

Doomsday Projections

It has been known for some time that, without strong measures to decouple economic growth from carbon emissions, a warming of 2 °C above pre-industrial levels could be surpassed by 2050 (IPCC 2018). The scenario of warming going to 4 °C is unthinkable with multiple dangers of far more extreme heatwaves, sharply rising sea levels, and far more severe storms, droughts, and floods. These have the gravest implications for the poorest and the most vulnerable segments of the population. Climate disasters have a disproportionate bearing on the poorer segments of the population (UNDRR 2022). Without climate-smart development and safety nets in place, climate change can push millions of people into extreme poverty by 2030 (The World Bank 2015).

The Amazon rightly gets enormous attention in the context of global prospects and catastrophes. It is vital to the global carbon cycle, being the largest tropical forest in the world. Higher temperatures dry vegetation and lead to droughts. These in turn debilitate trees and aggravate forest fires, releasing carbon and further warming the earth (UNEP 2022). Illegal logging, slash and burn agriculture, and most significantly, conversion of forest land for livestock production and plantations lead to massive deforestation, as mentioned in this chapter. More generally, deforestation has roots in macroeconomic policies of providing incentives for commercial agriculture, for example, subsidies for palm oil or soya cultivation as well as livestock, in place of protected land, generating "agricultural rents".

The melting of glaciers releases trapped carbon into the atmosphere, leading to further warming. The effect of carbon accumulation on floods and storms has been intensely studied. For example, one set of predictions is that a doubling of CO_2 concentrations in the air could triple the number of Category 5 storms (Anderson and Bausch 2006). As mentioned in Chapter 5, some envisage that for every 1 °C rise in global temperature, the frequency of events of the intensity of Hurricane Katrina would at least double (Grinsted et al. 2013).

Decimation of biodiversity is the biggest silent killer, every bit connected to human-made environmental destruction. The number of threatened species is increasing sharply, according to various sources. In one of them summarized in Fig. 6.1, the earth's biodiversity is under extreme threat, with an increasing number of endangered species every year. The pattern of economic growth and climate trends are related to the mistreatment of the environment and biodiversity. Nature is a precious asset that contributes to sustainable development and possesses intrinsic value (Dasgupta 2021; and the final chapter). The irreversible damage to biodiversity reflects the destructive path GDP growth is on. Protecting biodiversity is essential to ensure a degree of sustainability to development.

PROBLEMS ELUDING SOLUTIONS

The world has faced other daunting problems in recent decades: HIV/AIDS, COVID-19, the 2008 global financial crisis, for example.

Number of Animal Species of the IUCN Red List

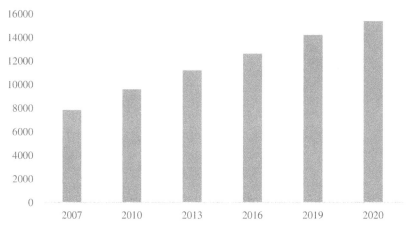

Fig. 6.1 Number of endangered species is rising (*Source* Buchholz 2021)

They all, as with climate change, are exceedingly dangerous, highly time sensitive, and tough to resolve. But climate change differs from the other crises in that it presents extreme conflicts of interest in finding solutions, for example, the fossil fuel lobby blocking low carbon pathways.

Degrees of Intractability

While also daunting, problems such as smoking, ozone layer depletion, and COVID-19, can be shown to have a clear line of direct causation. When the public connects cause and effect, responses are usually strong. For example, smoking has been established to have a direct contributory relationship with lung cancer. This has enabled effective policy responses (increased taxes on cigarettes, ban on public smoking, advertisements to discourage smoking) as well as changes in individual choices (Khang 2015). Furthermore, in the case of smoking and COVID-19, there is a direct personal identification that undergirds public opinion and the pressure to act (witnessing oneself or a family member getting sick with lung cancer or COVID-19). Air pollution can be identified intuitively and empirically as a contributing factor in respiratory illnesses. Climate change, however, is more indirectly related to human well-being. It

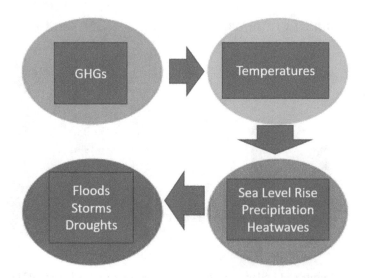

Fig. 6.2 The climate connection (*Source* Author's depiction; The Royal Society 2022)

involves steps in going from the source of the problem to its impact. GHG emissions raise the warming of air and water temperatures, increasing the chances of more droughts, heatwaves, and fires, on the one side, and worsening the intensity of storms and floods on the other. Warming causes sea levels to rise and increases the evaporation of water vapour, making for more extreme rainfall and floods, as well as adding greater energy to storms. These disasters, in turn, result in losses in lives and livelihoods. Figure 6.2 is a simplified depiction of the link between effluents and extreme weather.

Jointness and Collective Action

While climate change affects individuals, it is also a crisis requiring collective action. Shifting the trajectory of GHG emissions goes beyond the efforts of individuals or even countries. The solutions seen from a country's point of view call for others to respond as well. This means that answers to the question invoke concerns over free ridership, climate justice, and burden sharing. The inclination often is still to think that climate change is someone else's problem or to wait for others to step in and act.

Jointness of the needed response is present in other problems like a health calamity, but climate change presents it far more sharply, making its resolution that much more elusive. Xiang et al. (2019), in a study on climate change inaction, finds that the more intractable participants perceived climate change to be, the more demotivated they were to take climate action. This perceived intractability of climate change further induces individuals to remain unresponsive in the face of the crisis.

The need for collective action makes climate policy and investment subject to the effectiveness of coordination within and across borders. Since the mid-1990s, the UN COP summits have been yearly meetings on climate science, policy, and solutions. One of the most important bodies guiding the negotiations at COP is the IPCC, which is comprised of thousands of scientists and experts, and was established in 1988 by the WMO and the United Nations Environment Programme (UNEP). Despite the UN's leadership, concerns of free-riding and moral hazards have made coordination difficult. COP27 in 2022 takes place in Sharm El Sheikh, Egypt. The goals for COP27 include raising countries' ambitions on mitigation commitments to have a fighting chance to stay below the 1.5 °C mark. There need to be time-bound steps to phase out fossil fuels. The sense is that depending on the degree of implementation of the Glasgow COP 26 indications, the planet is headed towards 1.8 °C in the best case to 2.8 °C in the worst case, all well above the 1.5 °C mark (El Hatow 2022). Increasing the priority for climate adaptation is a second important area for decisions. A third area is establishing a global fund for "loss and damage" that poor nations battered by climate calamities can draw on. A fourth crucial objective of COP27 relates to public and private carbon financing which is discussed in Chapter 8.

Past outcomes of the COP summits have not been impressive. In highly volatile times of a downturn in the global economy, the challenge of joint action on climate change is harder. A breakthrough is needed in offsetting decades of lopsided emissions by rich countries. But they are unwilling to make up for past excesses by providing the massive financing that developing economies need for climate investment. Just as difficult is to stop the continuing and heavy burning of fossil fuels, for example by the top five emitters, China, the US, India, Russia, and Japan, as well as Southeast Asian countries taken together. Joint action could be signalled best by all nations announcing carbon neutrality and net zero by 2050 to help keep temperatures rise below 2 °C, if not 1.5 °C, as envisaged in the Paris Agreement. But this does not seem to be on the cards, especially in volatile times of economic hardships. There is merit in the COP summits to name the countries that are most out of line and urge them to do more.

There are instances where climate-friendly actions have been adopted, such as steps by individuals or companies for greater energy efficiency, or in making green investments, like green structures in urban areas, that have proven to be socially beneficial. But their scale remains modest and impacts inadequate. One aspect that deters society from taking meaningful measures on a sizeable scale is that the promised gains accrue over time or at a later date, although investments or costs must be incurred at the time of decision-making. The investments, therefore, fall short.

Personal Versus Societal Calculus

All this can be seen at the personal level. A case in point would be why people do not exercise or take preventive medication that they know will be good for them. One consideration that people seem to weigh is the cost of taking medication, which is felt immediately, whereas the benefits come later in time (List 2022). This could also explain COVID-19 vaccine hesitancy, in addition to the uncertainty over its effects. Underpinning this way of thinking is the bias in favour of the present that people have, that is, placing a higher value on immediate things. This issue of dealing with risks will be discussed in relation to discount rates in the next chapter.

Related is also the idea that people attach greater weight to losing something than to gaining an equivalent amount. This may be called a cognitive bias, explained in terms of loss aversion (Kahneman 2011). People prefer to avoid losses rather than pursue equivalent gains. The sensation of a loss seems to be greater than the appreciation of making a comparable gain. Such a benefit–cost calculus occurs at the personal as well as societal levels. In terms of climate change, at the societal level, the calculus based on immediate gratification is compounded by the ratio of future benefit and current costs being that much greater, given the nature of climate change as a slow onset problem which occurs at a global scale.

What is more, there is a cost to taking climate action, but the cost of not responding, by all estimates, is greater and rising. Wagner and Weitzman (2015) not only expresses this risk, but also acknowledges that short-term economic priorities could slow down the course of action. Indeed, the implied benefit of responding is a net win to society, but there are losses from the actions, at least in the short-term, to certain segments like the fossil fuel lobby (i.e., it is not all win–win). With alternative interests of winners and losers, some of the very champions of climate policy display divided loyalties, slowing reforms.

This observation draws attention to the fundamental political economy of reforms. All the scientific evidence and climate summits have not produced reforms that have moved the needle on climate outcomes. Effective change will need a good understanding of the politics of climate change and its historical antecedents (Mann and Wainwright 2020). As it looks increasingly unlikely that the planet will avert the less than 2 °C threshold for global warming, the political economy of the current directions, including the question of climate justice, assumes growing importance.

A Super Wicked Problem

Much has been written about the intractability of the climate crisis, and how it is a "super wicked" problem[1] (Levin et al. 2007, 2012; Lazarus 2010). Climate change is proving to be more intractable and difficult to address and hence more dangerous than extreme troubles like water pollution and epidemics.

Unholy Alliance

Super wicked problems could be understood to have four key features (Levin et al. 2007, 2012):

- Climate change is different from most social challenges because "the problem will, at some point, be too acute, have had too much impact, or be too late to stop or reverse" (Levin et al. 2007, 2012). Significant impacts will occur and, with each passing year, become more acute. If no action is taken, the risk of harm to human communities and ecosystems, as well as of non-linear changes and catastrophic events, increases.
- No coordinated global governance system exists that has proven effective in addressing global environmental problems like climate

change and enforce global emission limits, which underscores the lack of a central authority on this issue.
- There is also excessive discounting of the future that is irrational from the social point of view. This pushes responses in a way that undervalues the benefits of interventions that accrue in the future.
- Super wicked problems like climate change generate a situation in which the public and decision makers, even in the face of overwhelming evidence of the risks of significant or even catastrophic impacts from inaction, make decisions that disregard this information and reflect very short time horizons (Riedy 2013).

A Success Story

Some wicked problems, like smoking, ozone layer depletion and COVID-19, have seen coherent, effective global solutions. For example, the severity of the problem of ozone layer depletion has improved because of the 1987 Montreal Protocol on Substances that Deplete the Ozone Layer that resulted in a global response. From 1979 through the early 1990s, stratospheric ozone concentrations in the Southern Hemisphere fell to the concerning "ozone hole" level of 100 Dobson Units (DU). For several decades since the 1970s, concentrations continued to approximate 100 DU (Fig. 6.3). But since 2010 ozone concentrations have started to slowly recover, as data sources suggest. The ozone layer is expected to return to 1980 levels between 2045 and 2060 if all countries continue to meet their obligations and phase out the last ozone-depleting substances in the next few years.

The Montreal Protocol on Substances that Deplete the Ozone Layer aimed to ban the global production and use of ozone-damaging chemicals including CFCs, hydrochlorofluorocarbons, and halon. It broke new ground in its negotiation and in its construction. It is ratified or accepted by all 197 United Nations member states, a world first for any treaty and highlighting the strong global commitment to this treaty (Rae 2012). A winning formula then was that Dupont had a solution on the replacement of CFCs with HFCs and there were positive revenue gains. For the climate case, it is difficult to address the pervasive use of fossil fuels and the other GHGs in the same way.

6 INTRACTABILITY OF CLIMATE CHANGE 105

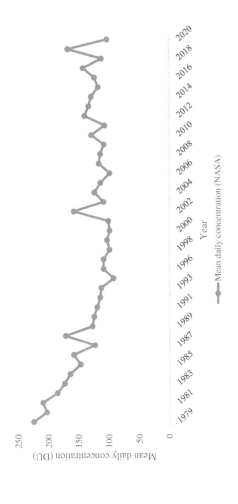

Fig. 6.3 Stratospheric Ozone Concentration (mean) (*Source* Our World in Data 2022)

Similarly, the efforts of the US and Canada to reduce acid rain that started in the 1990s are finally paying off (Smith 2015). A quarter of a century after emissions were reduced, soil acidification has been halted according to an interesting case study (see Smith 2015). Harmful aluminium in the soil (expressed as a ratio of recent to past values) clearly declined with greater reductions in atmospheric deposition of sulphate.

Compared to the progress on ozone, HIV/AIDs, and COVID-19, there are others that fall short on solutions despite characteristics that should make actions a matter of top priority. Air pollution is among the most vexing problems worldwide. There could be strong complementarities in addressing air pollutants and GHGs together as the sources of the problem in many instances are common to both effluents. The principal GHGs, as noted previously, include carbon dioxide (CO_2), methane (CH_4), nitrogen oxide (N_2O), ozone, and CFCs. The main air pollutants include particulate matter (PM), carbon monoxide (CO), sulphur dioxide (SO_2), nitrogen oxide, ozone, and lead. Air pollutants and GHGs are often released from the same sources. So, cutting GHGs, like CO_2, to slow climate change also reduces air pollutants, like particulate matter.

There are economies of scale in attacking the air pollution and climate change problems jointly. Transport, for example, is a major source of air pollution as well as GHGs. Delhi faces acute issues on both counts. Box 6.1 examines the pollution in Delhi and the role those vehicular emissions play in contributing to overall pollution levels.

Box 6.1 Delhi, Air Pollution and Transportation

Twenty-one of the world's 30 cities with the highest levels of air pollution are in India, according to IQAir AirVisual's *2019 World Air Quality Report*; moreover, 6 Indian cities are in the top 10 (IQAir 2021). Delhi has the dubious distinction of featuring the highest air pollution among the world's capital cities. Dacca and Baghdad are next in line among the capital cities.

Delhi's level of air pollution is extreme because of a combination of smoke from thermal plants and brick kilns in the capital region, effluents from a congested transportation network, stubble or biomass burning by farmers in neighbouring states, and the lack of cleansing winds that causes air pollution to hang over the city. The main sources of Delhi's particulate emissions are, in roughly equal measure, large powerplants and refineries, vehicles, and stubble burning. Even as technical solutions are within reach,

> the campaign must overcome the poor policy coordination among central, city, and local governments.
>
> Delhi needs a 65% reduction from the current baseline to meet even the national standards for $PM_{2.5}$, which are much more lenient than the World Health Organization (WHO) guidelines (CSE 2019). Effluent reduction needs to come from the major sources, a slew of industries, stubble burning in agriculture, and urban transportation. If vehicular emissions contribute roughly 18–39% of the $PM_{2.5}$ concentrations, tackling transport would be an important part of the solution. Benchmarked against comparable experiences, a one-third reduction from all sources by 2025 would be a goal to aim for.

LINES OF CAUSATION

It is common to hear references to hazards of nature as acts of God, as was the case in 2021 when a Himalayan glacier flooded the Indian state of Uttarakhand (Sengupta 2019). The extent of the glacier flooding was due to global warming. Furthermore, as glacier cover is replaced by water or land, the amount of light reflected decreases, aggravating warming—a contributor to the sweltering heat in Indian cities such as Delhi and Hyderabad.

The extreme cold weather in Texas, like the double-digit negative temperatures in Germany in early 2022, was connected to Arctic-peninsula warming mentioned earlier—at a rate almost twice the global average (WWF 2022). Usually there is a collection of winds around the Arctic keeping the cold locked far to the north. But with global warming, gaps seem to be appearing in these protective winds, allowing intensely cold air to escape and move south—a phenomenon that is accelerating (dw.com 2022). Reviewing both events show that the factors that precipitated their extreme natures were not natural; they resulted from human-made global warming. Unless climate change is clearly tagged as a primary culprit, there will not be decisive climate action.

Climbing global temperatures are also instigating greater use of air conditioning indoors, in turn driving up the global demand for electricity in the coming 30 years (Mellor 2022). More energy will need to be harvested, requiring the need for more fossil fuel burning, and thus contributing further to rising temperatures. Not only is this troubling economically and environmentally, but human lives are also put at risk due to the deadly nature of extreme heatwaves.

The indirect connection between fossil fuels and climate change is harder to think about amid a climate-related flood or heatwave, unlike the direct link between catching COVID-19 and falling ill. After all, GHGs are emitted by millions of businesses, which, over time, end up as accumulated concentrations in the air causing global warming and aggravating hazards. To inform people's awareness, it is essential to connect the emission of GHGs from the reliance on fossil fuel-based energy (as well as businesses like industrial farming and the food industry) as being partly responsible for extreme climatic conditions and the aggravation of hazards of nature (Kevany 2021). The section after the next points out the value of making this connection in real time when extreme weather events occur.

In the meantime, various analyses have been making crucial linkages. A study of the 2003 European heatwave and the droughts in winter in the Mediterranean point to human-induced climate change as being responsible for accentuating the chances of these hazards (Hoerling et al. 2012). Evidence also shows that GHG emissions contributed to the greater precipitation in two-thirds of the Northern Hemisphere regions (Min et al. 2011). Even precipitation extremes in short spells can prompt local flooding, soil erosion, and water damage. Studies on three catchment regions in south-eastern Australia suggest that scenarios that double CO_2 in the air can increase the frequency and size of floods with great damages to structures (Schreider et al. 2000).

Attribution reports have found that anthropogenic climate change made the floods that took 220 lives in Belgium and Germany in 2021 1.2 to 9.0 times more likely compared with when temperatures had not risen 1.2 °C from pre-industrial levels (Hodgson and Heal 2021). Scientists in the UK have attributed the July 2022 extreme heatwave to human activity, finding that the heatwave would have been 2–4 °C cooler if not for climate change caused by human activity (Rannard 2022). The probability of the extreme heatwave in India and Pakistan in 2022 is seen to be 30 times greater on account of climate change (World Weather Attribution 2022). Vautard et al. (2020) found that the 2019 European heatwave was about ten times more likely because of climate change. Similarly, the heatwave that hit the US in June was seen to be made 150 times more likely due to climate change (Lombrana, 2021).

Adding Fuel to the Fire

Global companies are often criticised for hiding the extent of the environmental damage they cause (Ambrose 2019). As the following examples suggest, the originators and perpetrators of the climate crisis are unlikely to disclose their responsibility for fear of losing business. The oil industry has misled the public on the fossil fuels-global warming relation by funding climate denial campaigns (Cook et al. 2019). Oil majors like ExxonMobil, from its own research, knew about the impact of the oil industry on GHGs for 40 years but covered up their contribution and misled the public (Hall 2015; Keane 2020).

The largest oil and gas companies have spent millions in blocking climate policies; buying influence in the EU is a case in point (Greenpeace European Unit 2019; Laville 2019). Organisations funded by corporations clearly show greater propensity to deny climate change (Warrick 2015). Ironically, there is an incentive for the big emitters to significantly raise their investments in mitigation, given their high exposure to weather disasters. The biggest carbon emitters—China, the US, India, Russia, Japan, and Germany—are responsible for 60% of total global emissions (Rapier 2019). Climate justice requires them to do the heavy lifting on climate investments.

That said, how carbon contributions of countries are counted does make a big difference. Table 6.1 taken from a recent study shows the tabulation of emissions as differentiated by the cumulative, the current total, and the current per person emissions. The rankings of countries vary according to which measure is used. The US is at the top in cumulative emissions, and China in the current aggregate.

It is important to distinguish between the stock and flow of emissions. Climate change depends on the stock or the concentration of GHGs in the atmosphere. What humans can control going forward as a response is a flow or the rate at which additional GHGs are emitted. If only the flow mattered, the damage could have dropped to zero if there is no emissions flow. But the extent of the damage depends on the stock of the pollutants. The flow of emissions has a positive lifespan, and they are being produced at a rate greater than the capacity of the environment to assimilate them.

Governments have enabled the use of polluting fuels by providing subsidies to producers and consumers. Worldwide, the International Energy Agency (IEA) estimates fossil fuel subsidies totalled US$300

Table 6.1 Emissions by Country

Country	Cumulative 1751–2014 (Gigatons CO_2)	% Global	Emissions 2014 (Gigatons CO_2)	% Global	Emissions per capita 2014 (tonnes CO_2)
China	174.7	12%	10.3	30%	7.5
United States	375.9	26%	5.3	16%	16.2
India	41.7	3%	2.2	6%	1.7
Russia	151.3	11%	1.7	5%	11.9
Japan	53.5	4%	1.2	4%	9.6
Germany	86.5	6%	0.7	2%	8.9
Iran	14.8	1%	0.6	2%	8.3
Saudi Arabia	12	1%	0.6	2%	19.5
South Korea	14	1%	0.6	2%	11.7
Canada	29.5	2%	0.5	1%	15.1
Brazil	12.9	1%	0.5	1%	2.6
South Africa	18.4	1%	0.5	1%	9.1
Mexico	17.5	1%	0.5	1%	3.8
Indonesia	11	1%	0.5	1%	1.8
United Kingdom	75.2	5%	0.4	1%	6.5
World	1434		34.1		4.7

Source Hsiang and Kopp (2018)

billion–US$500 billion a year over 2017–2019. But this financial estimate rises tenfold in IMF's calculation which includes the environmental damage from energy consumption (Carlin 2020; Coady et al. 2019). Countries also spend 20% more in financing fossil fuel projects (2019–2020) than on reducing pollution (Carrington 2021).

China is the largest public financier of overseas powerplants, oil, gas, and coal plants (Chen et al. 2020). Japan and the US are close behind. In 2020, nine multilateral development banks provided a combined US$3 billion to projects using fossil fuels—at odds with their stated support for the green economy (IISD 2021). Bilateral financiers, such as Japan Bank for International Cooperation (JBIC), and multilateral banks, such as the Asian Infrastructure Investment Bank (AIIB), Asian Development Bank (ADB), the World Bank, and others would want to refrain from any kind of explicit or implicit support for fossil fuels upstream or downstream, as the policy discussion in Chapter 8 suggests.

Many locations like the Pacific islands and parts of Africa are at the front end of climate dangers. Southeast Asia, McKinsey labels, is a region facing the greatest risks (Choudhury 2020). Already on the path of tropical storms originating from the Western Pacific and Indian Oceans, the region has seen a spike in disasters as global warming aggravates these hazards of nature. The dangers are compounded by the fact that the region also has a high population density, with large urban populations in low-lying cities, including the megacities, Jakarta, and Manila. Extreme weather has pushed intense storms and floods further inland because of rising sea levels, be it in Thailand, the Philippines, or Vietnam, and produced deadlier heatwaves in China and India.

At the same time the IMF finds the biggest jump in GHG effluents has been in Southeast Asia (Prakash 2018). High carbon financing to meet the region's surging energy demand aggravates climate change. The importance of finance goes further. It is both about eschewing the financing of fossil fuels and about encouraging investments in renewables. In the face of a global crisis, multilateral development banks have a quintessential role in promoting investments in clean sources of energy. In Asia, for example, there is a need to be far more ambitious in scaling up investments in renewable energy and supporting a low-carbon transition.

Coal is Best Left Underground

Coal presents the clearest line in the sand when it comes to what not to do. This concerns not only country governments and multilateral banks, but importantly, private businesses and their financiers. Coal is often mentioned as the lowest cost fuel. This is patently wrong. Inclusive of attributable health damages, the production and use of coal bear an extremely high social cost.

Unfortunatley, several countries are expanding their coal capacity. An ecologically troubling India-Australia partnership in coal mining and exports makes this point. The Indian conglomerate Adani was authorised to do a US$16.5 billion-dollar Carmichael Coal Mine project in Queensland, designed for massive exports of coal to India (Power 2017). Coal mining (and gas pipelines) provides income to the local economy and the workforce. The damage to health from burning coal in Australia could be on the order of US$2.6 billion annually (Climate Council 2019). But in densely populated India, coal contributes to 100,000 premature deaths annually (Goenka and Guttikunda 2020). Air pollution also accentuates death and illnesses from COVID-19. About 15% of deaths worldwide from COVID-19 could be attributed to exposure to air pollution (ESC 2020). Add to all these health damages the climate impact that spills over to all in Asia and the world.

This expansion of the coal enterprise in Australia comes at a time when social conscience and responsible business behaviour call for the converse. Seen in various settings worldwide, it is an example of private gain ahead of the social, and contrary to the need for sustainable business. On the other hand, a few countries have begun to take notice of the urgency to reign in fossil fuels and stopped financing coal.

Singapore's DBS Group, the Republic's largest bank, has decided to phase out funding coal (Ng 2021). Japan envisages no new coal plants (excluding ongoing construction), while JBIC has said it would stop financing new coal power projects. ADB has signalled the stopping of financing, which should apply to coal mining and upstream oil and gas exploration, drilling, and extraction as well as downstream activities (Nikkei Asia 2021). Meanwhile, financing of fossil fuels by China (and others), as well as AIIB, following its new energy strategy in 2022, would want to be reversed.

The major economies should also take the extra step to discourage trade in products that are high in carbon content. One way to do this is to impose a transnational carbon tariff on imports (OECD 2021). In Asia, the effect of this would be that China and India, as the major net exporters of carbon content, would pay higher tariffs, while Japan, Singapore, and Australia would be major net importers (Moran et al. 2018). All five economies can use some degree of monopsony power to tax the carbon content of imports—for example, Japan in machinery and equipment and Singapore in oil and gas.

A sea change is needed in how society views the role of fossil fuels in energy and development. Not only should emitters be made accountable but also enablers with special interests perpetuating them, through taxes and regulation. Because every loan portfolio has a carbon footprint, this content should be capped, with the cap being progressively made more stringent. For example, every pension fund's investment portfolio has a carbon footprint—this presents a great chance to tax investments in high carbon industries and activities. Similarly, controls on carbon emissions should be embedded in building codes, thus limiting their emission impact. Listed companies should be required to disclose their carbon footprints and this should be verifiable. Controversial as it is, it can be argued that carbon offsets can be helpful during a low-carbon transition. For example, if a fossil fuel producer builds a 1000-hectare forest, its positive effect could be considered an offset.

Enablers include not only governments and businesses, but also consumers in their expression of consumer preferences. The behaviour of consumers can play a big part in engineering a low-carbon transition of economies, ranging from decisions on low-carbon travel modes to food categories. Consumer attitudes could be nudged and incentivised to be less carbon-intensive through regulation where possible, and taxes.

An overhaul of global governance is needed to motivate governments to have a higher degree of accountability globally. Countries ought not to be free to destroy global public goods like how the Bolsonaro government in Brazil felt it was its sovereign right to burn down the Amazon. This calls for expanding the mandates of international financial institutions on these issues. There could be a need for a new international climate organisation staffed by climate scientists to follow through on verifying commitments governments make, rather than just leaving this to the bureaucrats.

Failure in Messaging

Reports outlining cause and effect tell a powerful, even if complex, story. They can motivate mitigation efforts, causally connecting fossil fuel-based energy and extreme weather events. It is also important to make the emission-disaster link when disasters strike and when public attention is most focused on disasters. Mainstream media is still not grasping the full significance of the climate-disaster link or making it clear in real time when it counts the most. In the wake of a string of natural hazards, there is greater labelling of climate change as a reality but the naming of human-caused climate change as the culprit is still occasional.

Climate connections are indeed complex, but scientists, economists, and the media need to do a better job in explaining them. On an issue with cataclysmic prospects, it is puzzling there is scepticism about its causes (Hoogendorn et al. 2020; Zaval et al. 2015). For far greater public awareness and a sense of urgency, the links in the chain need to be laid out in tangible and understandable ways, and this is best done in real time when disasters strike. There needs to be a clear identification of the causes of the predicament, responsibilities for the resulting disasters, and clarification of who should do what and when.

Climate events are headline news today. The climate-disaster link is also increasingly discussed. US President Joe Biden, during the visit to storm-ravaged New York and New Jersey in 2021 righty linked global warming to these catastrophes. That said, missing is the neglect of who really is to blame. Climate attribution reports can be enormously helpful in establishing cause and effect. Policymakers, economic advisors, the media, and weather reporters need to join the dots and inform public opinion.

When the UN's IPCC released a landmark report in 2018, warning that humanity had a mere 12 years to radically slash GHGs, only 22 of the 50 biggest newspapers in the US covered it (IPCC 2018; Macdonald 2018). In the US, news stories in 2019 about Hurricane Maria's devastation of Puerto Rico, and the epic floods in the Midwest hardly mentioned climate change in the live coverage (Hertsgaard and Pope 2019). As wildfires raged across the West Coast in 2020, cable news frequently sidestepped the relation between climate change and the fires, with only 13% of segments on the wildfires mentioning climate change (Monroe and Fisher 2020).

Moreover, when the top emitting nations display weak leadership in climate policy, the international response will remain weak (Leonhardt 2022). The politics of climate action is blocking progress with politicians going along with voters' reluctance to incur short-term costs that promise bigger gains. Campaigns by special interest groups supported by denials of climate science have an effect in shaping public opinion on priorities. A possible disenchantment towards climate science and scientists does not help the climate cause either (Krugman 2022).

Conflicts of interests imbedded in the climate predicament complicate the story. There needs to be straight talk about the ugly reality of powerful interests suppressing climate measures that benefit society. More generally, the interest in relying on cheap but polluting fuels is widespread. Llavador et al. (2015) argue that in all this the interest of economic growth of developing countries needs to be kept in mind. While developed countries pursue continued improvement of their societies' welfare, there needs to be a better collaboration among them on an increased focus on education and research in developing countries and aspirations for a differential growth rate.

In large part, this developed-developing country conundrum owes to the difficulty of policy coordination across countries, especially given the concerns of free-ridership and moral hazard. Despite rising climate-related disasters, there is little effective and coordinated global action towards tackling climate change. The ozone issue, which too is global, has faced challenges of coordination, free ridership, and moral hazard, but had some success in its resolution. It offers lessons on how polluters can end up acting in their as well as societal interest.

The discussion of the intractability of climate change leaves us with several questions. At a broad level, why is humanity not coming to grips with the reality of anthropogenic climate change? Is the lack of traction on climate responses due to causal linkages that are not intuitive and actionable? Could it be that it is more convenient to attend to more modest but immediately actionable issues and postpone intractable and slow-response ones, even though they are highly consequential?

Conclusions

An overarching conclusion of this chapter concerns the importance of establishing accountability if solutions to a seemingly intractable problem are to be found. While criticising lapses in disaster management related to climatic events, the ultimate liability must be placed on the promoters of fossil fuels, that exacerbate the warmer temperatures that are causing extreme hazards. The originators and perpetrators of the climate crisis, be they businesses or governments, need to be called out as they are unlikely to take responsibility that might cost them their businesses.

Science is clear that the use of fossil fuel-based, high carbon energy is causing atmospheric warming, in turn leading to extreme fires and floods. But public opinion is still not making a sufficiently strong climate-disaster link, and policy reforms lag. Policy responses in many countries linking air pollution to respiratory illnesses suggest that establishing cause and effect is key to making emitters accountable. Big emitters too can conceivably be seen to significantly raise their investments in mitigation, if the danger to them from disasters can become a decisive rationale. For now, this is not occurring on an adequate scale, with governments even making the use of polluting fuels more financially possible. A part of addressing the climate crisis is stopping the global financing of polluting fuels as well as government subsidies of fossil fuels, discussed further in the context of financing in Chapter 8.

Some countries and the MDBs have begun to take notice and restrict financing for coal. The major economies could also take the extra step of discouraging trade in carbon-intensive items such as steel and cement. Not only should emitters be made accountable but also enablers, through taxes and regulations. To reverse fossil fuel-based growth, policymakers, economic advisors, the media, and weather reporters need to communicate the linkages in the climate puzzle and press for the urgently needed changes.

NOTE

1. Levin et al. (2007) and Levin et al. (2012) build on Rittel and Weber's (1973) definition of wicked problems with four additional characteristics: that time is running out and it might soon be too late to stop or reverse damage, those wanting to end the problem are also causing it, there is no central decision-making authority, and that current policies discount the future irrationally.

BIBLIOGRAPHY

Ambrose, Jillian. 2019. "Major Global Firms Accused of Concealing Their Environmental Impact." *The Guardian*, June 16, 2019, sec. Environment. https://www.theguardian.com/environment/2019/jun/16/major-global-firms-accused-of-concealing-their-environmental-impact.

Anderson, Jason, and Camilla Bausch. 2006. "Climate Change and Natural Disasters: Scientific Evidence of a Possible Relation between Recent Natural Disasters and Climate Change." Think Tank. European Parliament.

Buchholz, Katharina. 2021. "Infographic: Number of Threatened Species Is Rising." *Statista Infographics*, August 4, 2021. http://www.statista.com/chart/17122/number-of-threatened-species-red-list/.

Carlin, David. 2020. "A 5 Trillion Dollar Subsidy: How We All Pay For Fossil Fuels." *Forbes*, June 2, 2020. https://www.forbes.com/sites/davidcarlin/2020/06/02/a-5-trillion-dollar-subsidy-how-we-all-pay-for-fossil-fuels/.

Carrington, Damien. 2021. "More Global Aid Goes to Fossil fuel Projects than Tackling Dirty Air—Study|Air Pollution." *The Guardian*, September 7, 2021. https://www.theguardian.com/environment/2021/sep/07/more-global-aid-goes-to-fossil-fuel-projects-than-tackling-dirty-air-study-pollution.

Centre for Science and Environment. 2019. "To Meet Air Quality Standards, Delhi Needs a 65 per cent Cut in Pollution Levels—Says New CSE Analysis." https://www.cseindia.org/to-meet-air-quality-standards-delhi-needs-a-65-per-cent-cut-in-pollution-levels-says-new-cse-analysis-9666.

Chen, X., Gallagher, Kevin P., and Denise L. Mauzerall. 2020. "Chinese Overseas Development Financing of Electric Power Generation: A Comparative Analysis." *One Earth* 3 (4): 491–503. https://doi.org/10.1016/j.oneear.2020.09.015.

Choudhury, Saheli Roy. 2020. "Southeast Asia Faces More Severe Effects of Climate Change than the Rest of the World, McKinsey Says." *CNBC*, August 16, 2020. https://www.cnbc.com/2020/08/17/southeast-asia-faces-more-severe-impacts-of-climate-change-mckinsey-says.html.

Climate Council. 2019. "Killer Coal: Just How Bad Are the Health Effects of Coal?" *Climate Council*, January 22, 2019. https://www.climatecouncil.org.au/killer-coal-just-how-bad-are-the-health-effects-of-coal/.

Coady, David, Ian Parry, Nghia-Piotr Le, and Baoping Shang. 2019. "Global Fossil Fuel Subsidies Remain Large: An Update Based on Country-Level Estimates." *IMF*, May 2, 2019. https://www.imf.org/en/Publications/WP/Issues/2019/05/02/Global-Fossil-Fuel-Subsidies-Remain-Large-An-Updated-Based-on-Country-Level-Estimates-46509.

Cook, John, Geoffrey Supran, Stephan Lewandowsky, Naomi Oreskes, and Ed Maibach. 2019. *America Misled: How the Fossil Fuel Industry Deliberately Misled Americans about Climate Change*. Center For Climate Change Communication. https://www.climatechangecommunication.org/america-misled/.

Dasgupta, Partha. 2021. *The Economics of Biodiversity: The Dasgupta Review: Full Report*. London: HM Treasury.

dw.com. 2022. "How Global Warming Can Cause Europe's Harsh Winter Weather|Environment|All Topics from Climate Change to Conservation|DW|11.02.2021." https://www.dw.com/en/cold-winter-global-warming-polar-vortex/a-56534450.

El Hatow, Lama. 2022. *Egypt Is Hosting COP27. What Are the expectations?* Atlantic Council. https://www.atlanticcouncil.org/blogs/menasource/egypt-is-hosting-cop27-what-are-the-expectations/

European Society of Cardiology. 2020. "Study Estimates Exposure to Air Pollution Increases COVID-19 Deaths by 15% Worldwide." https://www.escardio.org/The-ESC/Press-Office/Press-releases/study-estimates-exposure-to-air-pollution-increases-COVID-19-deaths-by-15-world, https://www.escardio.org/The-ESC/Press-Office/Press-releases/study-estimates-exposure-to-air-pollution-increases-COVID-19-deaths-by-15-world.

Fountain, Henry. 2019. "Climate Change Is Accelerating, Bringing World 'Dangerously Close' to Irreversible Change." *The New York Times*, December 4, 2019, sec. Climate. https://www.nytimes.com/2019/12/04/climate/climate-change-acceleration.html.

Goenka, Debi, and Sarah Guttikunda. 2020. *Coal Kills: An Assessment of Death and Disease caused by India's Dirtiest Energy Source*.

Urban Emissions. https://wayback.archive-it.org/9650/20200401013954/http:/p3-raw.greenpeace.org/india/Global/india/report/Coal_Kills.pdf.

Greenpeace European Unit. 2019. *Report: Big Oil and Gas Buying Influence in Brussels*. Greenpeace European Unit. https://www.greenpeace.org/eu-unit/issues/climate-energy/2238/big-oil-gas-buying-influence-brussels.

Grinsted, Aslak, John C. Moore, and Svetlana Jevrejeva. 2013. "Projected Atlantic Hurricane Surge Threat from Rising Temperatures." *Proceedings of the National Academy of Sciences* 110 (14): 5369–5373. https://doi.org/10.1073/pnas.1209980110.

Hall, Shannon. 2015. "Exxon Knew about Climate Change Almost 40 Years Ago." *Scientific American*, October 26, 2015. https://www.scientificamerican.com/article/exxon-knew-about-climate-change-almost-40-years-ago/.

Hertsgaard, Mark, and Kyle Pope. 2019. "The Media Are Complacent While the World Burns." *Columbia Journalism Review*, April 22, 2019. https://www.cjr.org/special_report/climate-change-media.php/.

Hodgson, Camilla, and Alexandra Heal. 2021. "Fires and Floods: Can Science Link Extreme Weather to Climate Change?" *The Financial Times*, August 5, 2021. https://www.ft.com/content/fe4e658e-0473-4f98-b995-4606aefa90bc.

Hoerling, Martin, Jon Eischeid, Judith Perlwitz, Xiaowei Quan, Tao Zhang, and Philip Pegion. 2012. *On the Increased Frequency of Mediterranean Drought in: Journal of Climate Volume 25 Issue 6 (2012)*. American Metereological Society. https://journals.ametsoc.org/view/journals/clim/25/6/jcli-d-11-00296.1.xml.

Hoogendoorn, Gea, Bernadette Sütterlin, and Michael Siegrist. 2020. "The climate change beliefs fallacy: The influence of climate change beliefs on the perceived consequences of climate change." *Journal of Risk Research* 23 (12): 1577–1589. https://doi.org/10.1080/13669877.2020.1749114.

Hsiang, Solomon, and Robert E. Kopp. 2018. "An Economist's Guide to Climate Change Science." *Journal of Economic Perspectives* 32 (4): 3–32. https://doi.org/10.1257/jep.32.4.3.

International Energy Agency. 2022. "Support for Fossil Fuels Almost Doubled in 2021, Slowing Progress toward International Climate Goals, According to New Analysis from OECD and IEA - News." *IEA*, August 29, 2022. https://www.iea.org/news/support-for-fossil-fuels-almost-doubled-in-2021-slowing-progress-toward-international-climate-goals-according-to-new-analysis-from-oecd-and-iea.

International Institute for Sustainable Development. 2021. *Fossil Finance from Multilateral Development Banks Reached USD 3 Billion in 2020, but Coal Excluded for the First Time Ever*. International Institute for Sustainable Development. https://www.iisd.org/articles/fossil-finance-multilateral-development-banks-reached-usd-3-billion-2020-coal-excluded.

Intergovernmental Panel on Climate Change. 2018. "Global Warming of 1.5°C." 2018. https://www.ipcc.ch/sr15/.
Intergovernmental Panel on Climate Change. 2022a. "Climate Change 2022a: Impacts, Adaptation and Vulnerability." https://www.ipcc.ch/report/ar6/wg2/downloads/report/IPCC_AR6_WGII_SummaryForPolicymakers.pdf.
Intergovernmental Panel on Climate Change. 2022b. "Climate Change: A Threat to Human Wellbeing and Health of the Planet: Taking Action Now Can Secure Our Future, Experts Say in New IPCC Report." *ScienceDaily*, February 28, 2022b. https://www.sciencedaily.com/releases/2022b/02/22022b8131507.htm.
IQAir. 2021. "World's Most Polluted Cities in 2021—PM2.5 Ranking|IQAir." https://www.iqair.com/in-en/world-most-polluted-cities.
Kahneman, Daniel. 2011. *Thinking, Fast and Slow*. New York: Farrar, Straus and Giroux.
Keane, Phoebe. 2020. "How the Oil Industry Made Us Doubt Climate Change." *BBC News*, September 20, 2020, https://www.bbc.com/news/stories-53640382.
Kevany, Sophie. 2021. "20 Meat and Dairy Firms Emit More Greenhouse Gas than Germany, Britain or France|Meat Industry|The Guardian." *The Guardian*, September 7, 2021. https://www.theguardian.com/environment/2021/sep/07/20-meat-and-dairy-firms-emit-more-greenhouse-gas-than-germany-britain-or-france.
Khang, Young-Ho. 2015. The Causality between Smoking and Lung Cancer among Groups and Individuals: Addressing Issues in Tobacco Litigation in South Korea. *Epidemiology and Health* 37: e2015026. https://doi.org/10.4178/epih/e2015026.
Krugman, Paul. 2022. "Opinion|Climate Politics Are Worse Than You Think." *The New York Times*, July 18, 2022. https://www.nytimes.com/2022/07/18/opinion/climate-politics-manchin.html?smid=em-share#after-story-ad-3.
Laville, Sandra. 2019. "Top Oil Firms Spending Millions Lobbying to Block Climate Change Policies, Says Report|Oil and Gas Companies|The Guardian." *The Guardian*, March 22, 2019. https://www.theguardian.com/business/2019/mar/22/top-oil-firms-spending-millions-lobbying-to-block-climate-change-policies-says-report.
Lazarus, Richard J. 2010. "Super Wicked Problems and Climate Change: Restraining the Present to Liberate the Future." *Scholarship @ Georgetown Law*, 84.
Leonhardt, David. 2022. "A Wilting Climate Response." *The New York Times*, July 18, 2022. https://www.nytimes.com/2022/07/18/briefing/extreme-heat-climate-fight-us-government.html.
Levin, Kelly, Benjamin Cashore, Steven Bernstein, and Graeme Auld. 2007. "Playing It Forward: Path Dependency, Progressive Incrementalism, and the

'Super Wicked' Problem of Global Climate Change". *In IOP Conference Series Earth and Environmental Science. Chicago: IOP Publishing.* https://doi.org/10.1088/1755-1307/6/0/502002.

Levin, Kelly, Benjamin Cashore, Steven Bernstein, and Graeme Auld. 2012. "Overcoming the Tragedy of Super Wicked Problems: Constraining Our Future Selves to Ameliorate Global Climate Change." *Policy Sciences* 45 (2): 123–152. https://doi.org/10.1007/s11077-012-9151-0.

List, John A. 2022. *The Voltage Effect: How to Make Good Ideas Great and Great Ideas Scale.* Currency.

Llavador, Humberto, John E. Roemer, and Joaquim Silvestre. 2015. "Sustainability for a Warming Planet." https://www.hup.harvard.edu/catalog.php?isbn=9780674744097.

Lombrana, Laura Millen. 2021. "Climate Change Made Germany Deadly Floods More Likely—Bloomberg." *Bloomberg Green*, August 23, 2021. https://www.bloomberg.com/news/articles/2021-08-23/climate-change-made-germany-deadly-floods-much-more-likely.

Macdonald, Ted. 2018. "Majority of Top U.S. Newspapers Fail to Mention Landmark Climate Change Report on Their Homepages." *Media Matters for America*, October 8, 2018. https://www.mediamatters.org/new-york-times/majority-top-us-newspapers-fail-mention-landmark-climate-change-report-their.

Mann, Geoff and Joel Wainwright. 2020, *Climate Leviathan: a Political Theory of Our Planetary Future.* Verso Books.

Mellor, Sophie. 2022. "Too Hot to Work? Calls for Maximum Indoor Working Temperature to Be Introduced as Heatwave Grips Europe." *Fortune*, July 12, 2022. https://fortune.com/2022/07/12/too-hot-to-work-tuc-calls-for-maximum-indoor-working-temperature-heatwave-europe/.

Min, Seung-Ki, Zhang, Xuebin, Francis W. Zwiers, and Gabriele C. Hegerl. 2011. Human Contribution to More-Intense Precipitation Extremes. *Nature* 470 (7334): 378–381. https://doi.org/10.1038/nature09763.

Monroe, Tyler, and Allison Fisher. 2020. "Cable News Largely Failed to Mention the Connection between Climate Change and Wildfires in Its Coverage." *Media Matters for America*, September 11. https://www.mediamatters.org/cable-news/cable-news-largely-failed-mention-connection-between-climate-change-and-wildfires-its.

Moran, Daniel, Ali Hasanbeigi, and Cecilia Springer. 2018. "The Carbon Loophole in Climate Policy: Quantifying the Embodied Carbon in Traded Products." The Carbon Loophole in Climate Policy, 65.

Newburger, Emma. 2021. "Disasters Caused $210 Billion in Damage in 2020, Showing Growing Cost of Climate Change." *CNBC*, January 7, 2021. https://www.cnbc.com/2021/01/07/climate-change-disasters-cause-210-billion-in-damage-in-2020.html.

National Oceanic and Atmospheric Administration. 2021. *Climate Change Indicators: U.S. and Global Temperature.* Reports and Assessments. https://www.epa.gov/climate-indicators/climate-change-indicators-us-and-global-temperature.
Ng, Kelly. 2021. "DBS to Exit Thermal Coal Lending by 2039." *The Straits Times,* April 16, 2021. https://www.straitstimes.com/business/banking/dbs-to-exit-thermal-coal-lending-by-2039.
Nikkei Asia. 2021. "Asian Development Bank to End Coal, Oil and Gas Financing—Nikkei Asia." https://asia.nikkei.com/Business/Energy/Asian-Development-Bank-to-end-coal-oil-and-gas-financing.
Organisation for Economic Co-operation and Development. 2021. "Carbon Dioxide Emissions Embodied in International Trade—OECD." https://www.oecd.org/sti/ind/carbondioxideemissionsembodiedininternationaltrade.htm.
Our World in Data. n.d. "Stratospheric Ozone Concentration." Our World in Data. https://ourworldindata.org/grapher/stratospheric-ozone-concentration.
Power, John. 2017. "India-Led Australian Megamine Goes Ahead despite Public Backlash - Nikkei Asia." *Nikkei Asia,* October 17, 2017. https://asia.nikkei.com/Economy/India-led-Australian-megamine-goes-ahead-despite-public-backlash2.
Prakash, Amit. 2018. "The Impact of Climate Change in Southeast Asia." *IMF Finance & Development Magazine,* September 2018. https://www.imf.org/external/pubs/ft/fandd/2018/09/southeast-asia-climate-change-and-greenhouse-gas-emissions-prakash.htm.
Rae, Ian. 2012. "Saving the Ozone Layer: Why the Montreal Protocol Worked." *The Conversation,* September 9, 2012. http://theconversation.com/saving-the-ozone-layer-why-the-montreal-protocol-worked-9249.
Rannard, Georgina. 2022. "40C Heatwave Has to Be Climate Change - Scientists." *BBC News,* July 29, 2022. https://www.bbc.com/news/science-environment-62335975.
Rapier, Robert. 2019. "The World's Top 10 Carbon Dioxide Emitters." *Forbes,* December 4, 2019. https://www.forbes.com/sites/rrapier/2019/12/04/the-worlds-top-10-carbon-dioxide-emitters/.
Riedy, Chris. 2013. "Climate Change Is a Super Wicked Problem." *Medium* (blog), May 29, 2013. https://medium.com/@chrisjriedy/climate-change-is-a-super-wicked-problem-b2e2b77d947d.
Ritchie, Hannah, Max Roser and Pablo Rosado. 2020. "CO_2 and Greenhouse Gas Emissions". OurWorldInData.org. https://ourworldindata.org/co2-and-other-greenhouse-gas-emissions

Rittel, Horst W. J., and Melvin M. Webber. 1973. "Dilemmas in a General Theory of Planning." *Policy Sciences* 4 (2): 155–169. https://doi.org/10.1007/BF01405730.

Schreider, S. Y., D. I. Smith, and A. J. Jakeman. 2000. *Climate Change Impacts on Urban Flooding.* Kluwer Academic Publishers.

Sengupta, Somini. 2019. "Rising Temperatures Ravage the Himalayas, Rapidly Shrinking Its Glaciers." *The New York Times*, 2019, sec. Climate. https://www.nytimes.com/2019/06/19/climate/global-warming-himalayas-glaciers.html.

Shabecoff, Philip. 1988. "Global Warming Has Begun, Expert Tells Senate." *The New York Times*, June 24, 1988. https://www.nytimes.com/1988/06/24/us/global-warming-has-begun-expert-tells-senate.html.

Smith, Kayla. 2015. "Soils Finally Recover after Decades of Acid Rain Limits|Great Lakes Echo." http://greatlakesecho.org/2015/12/16/soils-finally-recover-after-decades-of-acid-rain-limits/.

The Royal Society. 2022. "How Does Climate Change Affect the Strength and Frequency of Floods, Droughts, Hurricanes, and Tornadoes?" https://royalsociety.org/topics-policy/projects/climate-change-evidence-causes/question-13/

The World Bank. 2015. *Rapid, Climate-Informed Development Needed to Keep Climate Change from Pushing More than 100 Million People into Poverty by 2030.* World Bank https://www.worldbank.org/en/news/feature/2015/11/08/rapid-climate-informed-development-needed-to-keep-climate-change-from-pushing-more-than-100-million-people-into-poverty-by-2030.

Thompson, Clive. 2019. "How 19th Century Scientists Predicted Global Warming." *JSTOR Daily*, December 17, 2019. https://daily.jstor.org/how-19th-century-scientists-predicted-global-warming/.

United Nations Office for Disaster Risk Reduction. 2022. *Global Assessment Report on Disaster Risk Reduction.* https://www.undrr.org/publication/global-assessment-report-disaster-risk-reduction-2022

United Nations Environment Programme. 2022. *As Climate Changes, World Grapples with a Wildfire Crisis.* UNEP. http://www.unep.org/news-and-stories/story/climate-changes-world-grapples-wildfire.

Vautard, Robert *et al* 2020 *Environ. Res. Lett.* 15 094077 https://www.metoffice.gov.uk/research/climate/understanding-climate/attributing-extreme-weather-to-climate-change

Wagner, Gernot, and Martin L. Weitzman. 2015. *Climate Shock: The Economic Consequences of a Hotter Planet*, 1st ed. Princeton: Princeton University Press.

Warrick, Joby. 2015. "Why Are so Many Americans Skeptical about Climate Change? A Study Offers a Surprising Answer." *The Washington Post*,

November 23, 2015. https://www.washingtonpost.com/news/energy-env ironment/wp/2015/11/23/why-are-so-many-americans-skeptical-about-cli mate-change-a-study-offers-a-surprising-answer/.

World Meteorological Organization (WMO). 1994. *WMO Statement on the Status of the Global Climate in 1993*. Geneva: WMO. https://library.wmo. int/index.php?lvl=notice_display&id=7758#.Y0o-6XZBy3B.

World Meteorological Organization. 2022. "More bad news for the planet: greenhouse gas levels hit new highs". https://public.wmo.int/en/media/ press-release/more-bad-news-planet-greenhouse-gas-levels-hit-new-highs

World Weather Attribution. 2022. "Climate Chang Made Devastatingly Early Heat in India and Pakistan 30 times More Likely." May 2. https://www. worldweatherattribution.org/climate-change-made-devastating-early-heat-in-india-and-pakistan-30-times-more-likely/.

World Wildlife Fund. 2022. "Arctic Climate Change." https://arcticwwf.org/work/climate/.

Xiang, Peng, Haibo Zhang, Liuna Geng, Kexin Zhou, and Yuping Wu. 2019. "Frontiers|Individualist–Collectivist Differences in Climate Change Inaction: The Role of Perceived Intractability|Psychology." https://doi.org/10.3389/fpsyg.2019.00187/full.

Zaval, Lisa, Ezra M. Markowitz, and Elke U. Weber. 2015. "How Will I Be Remembered? Conserving the Environment for the Sake of One's Legacy." *Psychological Science* 26 (2): 231–236. https://doi.org/10.1177/095679761 4561266.

CHAPTER 7

A Persistently False Dichotomy

> No one would know from national statistics that natural capital is being degraded even as GDP is growing. Partha Dasgupta

While CO_2 emissions have been rising, economic-impact estimates, even since 2000, of the effects of climate change have largely been based on assumptions that ignore catastrophic changes, tipping points, and irreversibility (CO_2.earth 2022; NCEI 2019). Because of this, the resulting estimates of global economic losses from temperature increases beyond 2 °C seem quite low, between 0.2% and 2.0% of income (Keen et al. 2021). Even as one third of Pakistan goes under water with climate-driven epic floods and swathes of Europe suffer from the unbearable heat, some point to routine projections of GDP for 2100 as being hardly affected. Growth can be robust, but only if decoupled from emissions and damages.

Economists have not in the main anticipated the rapidly escalating climate crisis, just like they missed spotting a looming global financial crisis that broke in 2008. Some influential economists are starting to express concern over the threat global warming poses to sustainable economic growth. Even so, the traditional macroeconomic models used by governments and multilateral lenders underestimate or ignore climate effects in forecasting economic growth. As a result, governments have many incentives to boost short-term growth and far fewer to invest in mitigation. Furthermore, many governments have implicitly or

© The Author(s), under exclusive license to Springer Nature
Singapore Pte Ltd. 2023
V. Thomas, *Risk and Resilience in the Era of Climate Change*,
https://doi.org/10.1007/978-981-19-8621-5_7

explicitly supported high-carbon economic growth through investments or subsidies for investments. Even when mitigation efforts are taken, governments need to seize the opportunity to reduce inequities from climate impacts, given that climate measures and income distribution are intricately interlinked (Llavador et al. 2015).

CENTRALITY OF EXTERNALITIES

To begin changing this predicament, growth economics must factor in the priority for businesses and governments to tackle global warming. Fortunately, economists have the concepts and tools to do this in the form of externalities or spillover effects. They show that from individual projects to economy-wide programmes, counting these effects in the benefits and costs of economic activities and projects and programmes make a decisive difference in seeing the full picture and making the right choices. The examples range from global events like climate change to specific projects like protecting mangroves versus converting them into shrimp farms (Box 7.1).

Box 7.1 Mangroves or Shrimps?
Over the past two decades, Southeast Asia has replaced mangroves with shrimp farms, tourist resorts, and urban sprawl. These mangroves are commercially undervalued but provide valuable functions to society. Many species of fish depend on them as a nursery, while they are sites for nesting for birds. Mangroves are valuable for coastal protection, warding off soil erosion, and helping in adaptation to climate hazards. They are also a means for mitigation by serving as carbon sinks.

In Thailand, policymakers were faced with the decision to convert mangrove forests for shrimp farming or to leave them intact. Sathirathai (2001) did a valuation of mangroves by putting a price tag on their indirect services, namely, in coastal protection, as breeding habitat, and their ability to store carbon. Accounting just for marketed goods components, mangroves are valued at less per hectare than the potential per hectare from commercially farming shrimp. But, rightly adding in the role of mangroves in coastal protection, their value per hectare doubles, already supporting a decision not to convert. The mangroves' value per hectare rises further if its ability to store carbon, role as a nursery and breeding habitat for fisheries are accounted for.

> Mangroves worldwide are under severe threat and disappearing fast. Rising sea levels are severely hurting mangrove forests (Masters 2020). Unfortunately, by ignoring the social benefits of protecting mangroves or the negative externalities associated with converting them, mangroves in the upper Gulf of Thailand have been removed to make way for commercial shrimp farms and industrial development. Similar outcomes are evident in other parts of the world as well.
> *Source* Sathirathai (2001) and Lewis and Tietenberg (2019).

Pervasiveness of Spillover Effects

Social cost–benefit analyses, carried out at the country or global level, should ideally capture these externalities. Doing so makes a big difference to understanding the growth trajectories of countries and the influence of growth policies on socioeconomic outcomes (Quah 2015; Thomas and Chindarkar 2019). Key to obtaining this fuller understanding is the valuation of both tangible and intangible components of the social damage inherent in an investment, whether they were intended or not, for example, the health costs of air pollution generated by an infrastructure project.

Externalities can spring from the consumption of goods and services, like emitting pollution by driving cars using carbon-intensive gasoline (Perman et al. 2003). They can also be caused by the production of goods and services, like a chemical plant polluting the waterways. Or they can be a combination of both. One can also imagine that any of the three above-mentioned sources is directed at consumers or at producers or at both. All these distinctions can be applied to both positive and negative externalities which are not being internalised or compensated for. Spillovers can be positive, as when a vaccination programme provides secondary health benefits to communities. But climate change involves knock-on damages because firms do not account (or pay) for the societal harm caused by their operations—such as a power plant emitting carbon dioxide, which contributes to global warming, whose effects are felt globally.

If a project is funded by a national government but benefits a local area, a cost–benefit analysis would look different depending on whether it is viewed from the national or local viewpoint. It is vital that the social benefits are measured even if difficult to do. Environmental projects usually trigger both direct and indirect consequences. Tangible benefits are those that can reasonably be assigned a monetary value; intangible benefits are those that are difficult to quantify, either because data are not available or reliable enough or because it is not clear how to measure the value even with data.

Cost estimates of pollution control can be requested from cost-bearers. But the problem with this approach is the strong incentive for cost-bearers to overstate or understate the estimates depending on expectations about the use of the information. Another is an approach using engineering information to catalogue technologies to estimate the costs of using those technologies. Analysts frequently use a combination of survey and engineering approaches.

In cost–benefit analysis, the choice of the discount rate, using which future benefits and costs are adjusted to reflect people's preference for the current over the future, is very significant. Heal (2017) points out the different conclusions that are reached on the need for mitigation to reduce GHG emissions when alternative discount rates and damage functions are specified. The long-term nature of climate change gives rise to extreme sensitivity in determining discount rates. Integrated assessment models (IAMs) provide a means to understand relationships in climatic action systems. But discount rate uncertainties, coupled with tenuous predictions of the impacts of mitigation, render the conclusions using IAMs inaccurate for drawing policy implications (Stern and Stiglitz 2022).

Discount rates reflect rates of time preference. A divergence in time preferences can cause not only a wedge between private and social discount rates (as when firms have a higher rate of time preference than the public sector), but even between similar analyses in different countries. Time preferences are sometimes considered to have a greater weighting in a cash-poor, developing country than in an industrialised country. And when cost–benefit analyses in two countries are based on different discount rates, they can arrive at quite different conclusions. Uncertainty over impact continues to limit the usefulness of IAMs (Heal 2017).

Growth Economics and Policy

Governments give incentives to boost short-term GDP growth even when these incentives are based on high-carbon infrastructure, which aggravates environmental problems. Targets for economic growth need to shift the focus from the concept of gross domestic product to net domestic product that is inclusive of social and environmental impacts. Short-term economic forecasts as well as longer-term projections are still preoccupied with slight changes in GDP growth rate, failing to integrate the far bigger concern over climate change derailing growth itself. Calls have been made to eliminate negative externalities, and to compensate for the remaining pollution (Elliott and Esty 2021).

Externalities are not organically embedded in research and teaching in growth economics—and, as such, growth models can often be used to support extreme deregulation, improving the business environment for enterprises at the expense of the environment and society. Government policies often neglect these implications, but sometimes go further and accentuate the harm. For example, the policies of the Bolsonaro administration in the Amazon during 2019–2022 have encouraged the deforestation of the tropical rainforest, contributing to carbon dioxide emissions. US policy under the Trump administration represented an alarming failure to consider the climate crisis. By replacing the previous administration's modest Clean Power Plan with a far weaker Affordable Clean Energy rule, and promoting more coal-based energy, it encouraged higher CO_2 emissions (Irfan 2019). As a result, US emissions from energy sources alone increased by 2.7% from 2017 to 2018 and were 4.5% higher in the first three months of 2019 than in the same period the year before (EIA 2022).

These deleterious effects have received plenty of attention in the literature, but there is yet another mostly overlooked factor that is decidedly in the way of environmental progress: the mainstream economics profession has not been on board in the campaign for climate action. History has shown that economists are influential actors in creating the necessary consensus for far-reaching policy reforms. To see the potential of economics, consider how it has aided in translating intangible policy matters into everyday concerns of people, in turn spurring policy responses. The economics of investing in girls' education or strengthening primary health care are cases in point.

There have been discussions on climate and development (IMF 2019). But growth economics is largely silent on climate change or worse, applauds growth even when it is mostly produced by burning fossil fuels. Some indicators of economic performance ignore these harmful effects, giving high rankings to countries on growth or competitiveness even though their outcomes on environmental and social sustainability, which are integral to economic progress, are negative. The World Bank's widely used Doing Business indicator which has been discontinued for the time being, was a prime example of applyiing faulty economics and causing extensive socio-environmental damages (Picciotto and Thomas 2021).

One of the reasons for the disinclination of economists to wade into the climate problem is the lack of methodological tools for analysis, as well as the frame of mind in the teaching of mainstream economics, for the study of climatic impacts. This is evident when influential economic journals and media channels rarely feature environmental and climate implications for economic growth and well-being. A review of a 2019-published article by Oswald and Stern found that the widely cited *Quarterly Journal of Economics* had not published a single article on climate change and the quantitative journal, *Econometrica* only two (Oswald and Stern 2019). Overall, this puzzling picture has improved a bit since that assessment (McLaughlin 2021).

Scientists, for their part, have not often been open to integrating the work of economists into climate research. One understandable reason for this is that some of the premises on which mainstream economists build their models are indefensible—for example, the single-minded focus on the quantity of growth, while neglecting its quality in terms of environmental and social sustainability (Thomas et al. 2000). But often scientists and science journals do not seem to want to recognise the economic analysis of climate change, presumably because of the stark differences in methodologies used by economists. They also do not focus much on the valuation of intangible and indirect impacts.

Economists have nevertheless begun to make valuable contributions to solving the climate crisis, as indicated earlier, and several policy institutes have been incorporating climate realities into economic analysis and policy advice. A major public statement was issued to support market-based responses to climate change. The *Economists' Statement on Climate Change* was published in 1997, just before that year's Kyoto Protocol flagged the climate danger and proposed market-based solutions (for example, Jorgenson 1997). The Centre for Social and Economic Progress based in New Delhi, for instance, puts special emphasis on climate economics in its policy work (Ahluwalia and Patel 2022). The Grantham Research Institute for Climate Change and the Environment based in London has done work on the interface of natural science and economic policy (Muller and Robins 2022). A few influential bodies in the financial world, including the Network for Greening the Financial System, a group of central banks, are trying to raise the world's consciousness about climate risks.

Discounting the Future

The agreed warming limit of the 2015 Paris Agreement would seem to have influenced countries in valuing the future, demonstrated with an increasing number of nations establishing a target of carbon neutrality by 2050. Climate regulation would seem to prompt announcements of climate investments. A true valuation of the future, however, hinges on far-reaching structural changes in economies to achieve the global warming targets.

Furthermore, the scientific consensus has been at odds with a major stream of thought within the economics profession. As mentioned earlier, a great deal of the economics of climatic change has centred on IAMs. These models can and do incorporate different economic and social elements, but by not reflecting the scientific literature adequately, they generate highly problematic results. Accordingly, a mainstream view is that "societal optimisation" entails accepting an increase in temperature of some 2.5–4 °C (Keen 2019). This is an increase rightly seen as catastrophic by many, especially climate scientists (Stern et al. 2021).

Climate work has been recognised. The IPCC and Al Gore, former US Vice President, were awarded the Nobel Peace Prize in 2007 for their work on "man-made climate change" (UN 2007). William Nordhaus won the Prize in Economics in 2018 for "integrating climate change into long-run macroeconomic analysis" (YaleNews 2018). But surprisingly, this work and its follow up seriously understated the urgency for climate investment by not recognising the science of immense damages, tipping points, and irreversibility of certain losses—which has seriously hurt the advancement of the climate agenda (Nordhaus 2007, 2017; Thomas 2019; Keen 2020).

This work, in focusing on the economics of growth, heavily discounts the gains from climate mitigation and overestimates the costs of climate measures. It does not recognise that climate impacts involve rapidly accumulating ecological damage that is pervasive and sometimes irreversible. Thus, ironically, when some leading economists have weighed in on climate change, showing the profession's dominant influence on public political debate, their mistaken line of argument has been in the way of taking timely climate action. It is not clear whether these economic arguments were in part the cause or a later justification of the slow pace of climate reform.

> **Box 7.2 Climate Change and Prices**
> In discussing spillover effects and investment with lags in their effects, prices and interest rates come into play. Consider Table 7.1. The upper-left cell shows transactions based on market prices and market interest rates, without counting externalities or future effects. The upper right cell presents the effects of price signals that correct for this failure, for example, via carbon taxes. While decisions made on this basis do account for externalities, they do so only from the viewpoint of "normal" investors and their time frames. The upper right cell corresponds to Nordhaus' picture of heavily discounting long-term climate investments.
>
> The lower two cells introduce corrections to interest rates, treating the welfare of future generations as being like that of the current. The lower left cell, while reflecting market prices, uses normative rather than market interest rates. This accommodates the interests of future generations, while some argue that this leads to wasteful subsidies even to ordinary investments. Moving to the lower right corner introduces price signals that reflect externalities as well as the interests of future generations. Decisions made on its basis try to correct not only for current cross effects but also inter-generational ones. The lower right cell corresponds to the mitigation strategies in the Stern review (Stern 2007).

If negative externalities were central to economic policymaking, policymakers would apply carbon taxes far more widely and intensely to shape policy. Country experiences point out the value of economic policy recognising their central role in the growth process and integrating the costs and benefits of environmental stewardship and climate investment. By giving weight to these cross effects, economics could make an enormous contribution to tackling global warming.

Many economists, Nordhaus among them, support the introduction of a carbon tax. More than 40 countries, including China where a carbon

Table 7.1 Prices, interest rates and externalities

Price discounting	Market prices	Externality-corrected prices
Market interest rates	Market decisions	Decisions that internalise the external effects of transactions
Normative (low) interest rates	Decisions that internalise the interests of future generations	Decisions that internalise external effects and the interests of future generations

Source Petri and Thomas (2013)

market became fully operational in 2021, now use some form of carbon-pricing, a carbon tax, or emission trading (The Economist 2019; Black et al. 2022). But these measures still cover only about one-half of these countries' emissions, representing some 13% of the global total. In its origins and solutions, the climate crisis is intrinsically socioeconomic and political in nature. The way forward is to benefit from the economic analysis and tools to solve the physical manifestations. The need of the hour, therefore, is a greater application of economic analysis that integrally complements scientific research.

Economists should focus on establishing a causal link between carbon accumulation in the atmosphere and the observed spike in natural, social, and health disasters. Using econometric and other methods of economic analysis, economists could considerably improve the receptivity of policymakers, many of them social scientists, to climate solutions.

ECONOMETRICS AND CLIMATE DISASTERS

Econometric and other methods in social sciences are broadening scientific findings geographically as well as thematically. The use of massive panel data comprising most countries over a long period of time has enabled a wider scope. The use of econometric techniques is not new to the climate change literature (Juselius 2007; Kaufmann et al. 2011; Kaufmann and Stern 2002). But the inclusion of climatic factors together with socioeconomic considerations in the same framework is relatively new, and more so at the global level.

Emissions and Disasters

López et al. (2016) shows that atmospheric carbon dioxide accumulation significantly increases hydro-meteorological disasters, which in turn induce significant negative effects on the rate of economic growth. Lopez et al. (2020) provide a framework for including socioeconomic and demographic factors, as well as climate-related ones that affect the incidence hydro-meteorological disasters. As in this work, it is useful for policy to assess the increases in the probability of disasters on account of carbon accumulation and global warming, after recognising the difference exposure and vulnerability make.

Establishing a direct causal relationship (for example, ozone layer depletion and skin cancer) with a personal identification helps to spur a global response. Showing long-term relationships, while useful, is not enough to change public opinion; there must be a demonstration of cause and effect and accountability—that is the significance of the new regional and global climate assessments.

The value of these studies lies in gathering the evidence of effects beyond country level ones to shed light on global impacts and to press the need for collective climate action. Recent research examines the connection between carbon dioxide concentrations in the air and the probability of hydro-meteorological disasters using almost all countries since 1970 (Fortunato et al. 2022). A key finding of this work is that the probability of a disaster is comprised of the effects of country-specific and global factors. This separates out the impact of global changes in atmospheric emissions arising from multiple sources from the local factors that affect countries very differently. Indonesia is more susceptible than Singapore by virtue of its greater exposure to hazards of nature, but both are hit by the global force of the carbon build-up in the air.

This line of research also evaluates the climate impact within a framework that includes other key variables, exposure, and vulnerability that can turn hazards of nature into disasters. Exposure to climate effects has socioeconomic origins, such as population density in coastal areas, as does vulnerability to these potential dangers; for example, poverty and the capacity to cope with it. The probabilities of hydro-meteorological disasters at country levels have been estimated, while assigning a part of the outcomes to anthropogenic climate change. This line of work adds urgency to find climate solutions.

Policy Influence

Some studies show the extent to which people's exposure and their vulnerability, in addition to the carbon-intensity of activities, turn extreme hazards of nature into human disasters. For example, the impact of extreme events worsens when there is a greater density of population and a higher degree of poverty. The question here is, after having accounted for these factors, how much of the extreme events can be attributed to the continuous rise in atmospheric CO_2 concentration and global warming.

Lopez, Thomas, and Troncoso (2020) suggest that if the CO_2 level increases by 1%, floods and storms would likely increase by nearly 9%. The yearly increase in CO_2 has been about 2.4 ppm or about 0.6% from the base 396.5 ppm level for 2010–16. Accordingly, the number of intense hydro-meteorological disasters could increase by 5.4% annually for an "average" country facing annually nearly one extreme disaster (defined as one that causes 100 or more fatalities or affects 1000 or more people). So, with the current trend in CO_2 accumulation, the number of intense floods and storms could double for this average country (i.e., one more extreme event) in a span of 13 years.

Fortunato et al. (2022) applies a cointegration analysis with CO_2 concentration to the time paths of warming and disasters and provide projections of the probability of disasters by country to the year 2040. It finds a stable relation between CO_2 accumulation and disasters that allows projections of the latter being conditional on the former. This statistically significant predictor of the probability of occurrence of hydro-meteorological disasters for a specific country could be a basis for policymakers to assess the dangers ahead of time. The probability of a country disaster is divided into the effects of country-specific factors, such as climatic and socio-demographic factors, and factors associated with the global climate. This work detects a stable relation between CO_2 accumulation and the global climate that allows a projection of the latter process as conditional on the former. Projections like these could feed into efforts to motivate decarbonisation.

But at the individual and even country level, the motivation to do the opposite is powerful. As mentioned in Chapter 5 looking ahead, there are also potential non-converging spirals, rather than self-correcting forces, inherent in the climate conundrum whose timely treatment calls for redoubled analytical efforts. One example of this is the greater use of cooling, like air conditioning, in response to global warming. Another is the energy crunch in the wake of supply bottlenecks caused by global warming, putting pressure on ramping up fossil fuel-based energy and exacerbating global warming. Texas' widespread electricity failure in 2021 was largely caused by the freezing of natural gas pipelines due to weather extremes (Searcey 2021). This and other energy shortage related to global warming ironically led to more reliance on fossil fuels. Uncertainty over energy supplies could motivate a race to the bottom, creating a vicious cycle between increasing economic uncertainty and pollution. It is therefore more important that the evidence of the damaging impacts of human-made climate change is shared and understood.

The onset of COVID-19 and lockdowns and mobility restrictions of 2020 was a breather for the relentless rise in global carbon emissions. But this respite was short lived and the uptick in effluents resumed in 2021, the year nearly 200 countries gathered in Glasgow for COP 26. In the case of the US, emissions rose 6% in 2021 after a record 10% decline in 2020, as the economy rebounded, fuelled by coal power and truck traffic (Plumer 2022). Continuing on this track will lead to scenarios of an ice-free planet in the coming decades (Borunda 2020).

Economics at the Climate Table

The question must be asked why a good segment of the economics profession, which is highly influential when it comes to policy making, has chosen to ignore the climate concern over the decades. Indeed, much of the economics of growth practically encourages any type of increase in GDP, even those that come at the cost of environmental damage. Mainstream economics, wedded to a tradition of finding means in its toolkit to maximise economic growth, has not found environmental action convenient. This is changing as evidence on climate damages mounts (Carton and Natal 2022). Scientists, on their part, seem reluctant to interact with the economics profession on the incipient climate economics.

The scientific community rightly finds some of the premises of mainstream economics, such as its neglect of ecological damage where science has weighed in, unacceptable. That said, and following Chapter 6, if scientific investigation were to dovetail with economic analysis, there would be considerable gains in terms of a better understanding of the issues, formulation of policy responses, as well as communication of the problem.

By integrating case studies from scientific investigation with the findings of economic analysis, scientific results of the physical phenomena can be enriched, bringing to life climate scenarios and policy responses. The recent use of economic analysis in climate change studies is promising. It appears that the scientific findings and those obtained by studies using economic methodologies are quite consistent, each providing different perspectives of the problem. This consistency of findings of the two approaches using very different methods strengthens the confidence in the results. This variety of perspectives reaching similar conclusions also provides insights into the problem that could enable the results to be appreciated by a broader audience, including economists who have hitherto been sceptical about the efficacy of climate policies.

That said, the teaching and practice of economics have become diverse, especially with the emergence of many who want to find interdisciplinary solutions. So, scientists must update their impression of economists, and allow synchronisation between science and economics on climate policy. Such alignment will also require a new mindset across the economics discipline—from teaching at schools and universities to practice in economic policymaking.

Conclusions

Economics needs to be agile in recognising the rapidly worsening climate risk. The false separation between environmental protection and economic growth needs to give way to holistically consider sustainable development. Economic policy advice needs to give central attention to the spillover harm to society from economic activity. This damage also needs to be integrated into projections of economic growth, policy advice, and project analysis.

Economic analysis needs to stop encouraging GDP growth that does not deduct the social and environmental damage. Macroeconomic models used by governments and their development partners, including the IMF, World Bank, and other multilateral lenders, have, by and large, underestimated or ignored climate effects in forecasting growth, often on account of the inability of these models to capture environmental effects. As a result, many governments have continued to use many incentives to boost short-term GDP growth, while doing relatively little to invest in environmental protection and climate mitigation, which generally take time to yield visible results.

There is a great need for cost–benefit analysis of development initiatives, carried out at the level of a country or globally, ideally capturing the spillages, both positive and negative. The analysis must be about the social calculus and not just the private. In the case of climate change, the dominant impacts are negative. Such externalities need to be seen as central to economic analysis, prompting policymakers to apply instruments like carbon taxes far more widely and intensely to guide economic growth. Economics also needs to get away from the unhelpful premise that growth will make future generations rich enough to redress climate risks, in turn justifying the heavy discounting of the benefits of climate investment in future years. The use of discount rates in climate analysis should be calibrated to be fit for purpose.

A change is needed across the economics discipline, at schools and in policymaking; in fact a wholesale change would be welcome to encourage interdisciplinary treatments. Economics can be very insightful on the causes and solutions to the climate crisis, which are inherently socioeconomic in nature. And science provides the physical basis for understanding the phenomenon. Scientific breakthroughs have driven solutions to some of the toughest problems. Of far-reaching significance would be a coupling of science and economics to act on influencing climatic trends. Economics would need to integrate scientific evidence on irreversible ecological damage.

Bibliography

Ahluwalia M.S. and U. Patel. 2022. *Climate Change Policies for Developing Countries*. New Delhi: Centre for Social and Economic Policy. https://csep.org/working-paper/climate-change-policy-for-developing-countries/.

Black, Simon, Ian Parry, and Karlygash Zhunussova. 2022. "More Countries Are Pricing Carbon, but Emissions Are Still Too Cheap." *IMF Blog* (blog), July 21, 2022. https://blogs.imf.org/2022/07/21/more-countries-are-pricing-carbon-but-emissions-are-still-too-cheap/.

Borunda, Alejandra. 2020. "Arctic Summer Sea Ice Could Be Gone by as Early as 2035." *Science*, August 13, 2020. https://www.nationalgeographic.com/science/article/arctic-summer-sea-ice-could-be-gone-by-2035.

Carton, Benjamin and Jean-Marc Natal. 2022. "Further Delaying Climate Policies Will Hurt Economic Growth." *IMF Blog*. October 5. https://www.imf.org/en/Blogs/Articles/2022/10/05/further-delaying-climate-policies-will-hurt-economic-growth.

CO_2.earth. 2022. "Daily CO_2." https://www.CO2.earth/daily-CO2.

EIA. 2022. "Total Energy Monthly Data—U.S. Energy Information Administration (EIA)." https://www.eia.gov/totalenergy/data/monthly/.

Elliott, E. Donald and Daniel C. Esty. 2021. "The End of Environmental Externalities Manifesto: A Rights Based Foundation for Environmental Law." *NYU Environmental Law Journal* 29. January 7.

Fortunato, Andrés, Helmut Herwartz, Ramón López, and Eugenio Figueroa. 2022. "Carbon Dioxide Atmospheric Concentration and Hydrometeorological Disasters." *Natural Hazards*, January, 1–18. https://doi.org/10.1007/s11069-021-05172-z.

Heal, Geoffrey. 2017. "The Economics of the Climate." *Journal of Economic Literature* 55 (3): 1046–1063. https://doi.org/10.1257/jel.20151335.

International Monetary Fund. 2019. "The Economics of Climate." Finance and Development. https://www.imf.org/external/pubs/ft/fandd/2019/12/pdf/fd1219.pdf.

Irfan, Umair. 2019. "Trump's EPA Just Replaced Obama's Signature Climate Policy with a Much Weaker Rule." *Vox*, June 19, 2019. https://www.vox.com/2019/6/19/18684054/climate-change-clean-power-plan-repeal-affordable-emissions.

Jorgenson, Dale W. 1997. "The Economics of Climate Change." https://www.epw.senate.gov/105th/jorg0710.htm.

Juselius, Katarina. 2007. "Cointegration Analysis of Climate Change: An Exposition," August, 39.

Kaufmann, Robert K., Heikki Kauppi, Michael L. Mann, and James H. Stock. 2011. "Reconciling Anthropogenic Climate Change with Observed Temperature 1998–2008." *Proceedings of the National Academy of Sciences* 108 (29): 11790–11793. https://doi.org/10.1073/pnas.1102467108.

Kaufmann, Robert K., and David I. Stern. 2002. "Cointegration Analysis of Hemispheric Temperature Relations—Kaufmann—2002—Journal of Geophysical Research: Atmospheres—Wiley Online Library," *Journal of Geophysical Research*, 107 (D2). https://doi.org/10.1029/2000JD000174.

Keen, Steve. 2019. "'4°C of Global Warming Is Optimal'—Even Nobel Prize Winners Are Getting Things Catastrophically Wrong." *The Conversation*. http://theconversation.com/4-c-of-global-warming-is-optimal-even-nobel-prize-winners-are-getting-things-catastrophically-wrong-125802.

Keen, Steve. 2020. "The Appalingly Bad Neocassical Economics of Climate Change." *Globalizations* 18 (7). September 1. https://www.tandfonline.com/doi/abs/10.1080/14747731.2020.1807856.

Keen, Stephen, Timothy M. Lenton, Antoine Godin, Devrim Yilmaz, Matheus Grasselli, and Timothy J. Garrett. 2021. "Economists' Erroneous Estimates of Damages from Climate Change." http://arxiv.org/abs/2108.07847.

Lewis, Lynne, and Thomas H. Tietenberg. 2019. *Natural Resource Economics: The Essentials*. Routledge.

Llavador, Humberto, John E. Roemer, and Joaquim Silvestre. 2015. "Sustainability for a Warming Planet." 2015. https://www.hup.harvard.edu/catalog.php?isbn=9780674744097.

López, Ramón E, Vinod Thomas, and Pablo A Troncoso. 2016. "Is Climate Change behind the Rise in Natural Disasters?," 12.

Lopez, Ramon E., Vinod Thomas, and Pablo A. Troncoso. 2020. "Impacts of Carbon Dioxide Emissions on Global Intense Hydrometeorological Disasters." January 7, 2020. https://doi.org/10.18783/cddj.v004.i01.a03.

Masters, Jeff. 2020. "Sealevel Rise Likely to Swallow Many Coastal Mangrove Forests." *Yale Climate Connections*. June 10. https://yaleclimateconnections.org/2020/06/sea-level-rise-likely-to-swallow-many-coastal-mangrove-forests/.

McLaughlin, Eoin. 2021. "How Have Economists Thought about Climate Change?" *Economics Observatory*, October 28, 2021. https://www.economicsobservatory.com/how-have-economists-thought-about-climate-change.

Muller, Sabrina and Nick Robbins. 2022. Just Nature: How finance can support a just transition at the interface of action on climate and biodiversity. *GRICCE.LSE*. https://www.lse.ac.uk/granthaminstitute/publication/just-nature-finance-just-transition-climate-and-biodiversity-2022/.

National Centers for Environmental Information. 2019. "Assessing the Global Climate in May 2019|News|National Centers for Environmental Information (NCEI)." https://www.ncei.noaa.gov/news/global-climate-201905.

Nordhaus, William, D. 2007. "A Review of the Stern Review on the Economics of Climate Change." *Journal of Economic Literature*, 45 (3): 686–702. https://doi.org/10.1257/jel.45.3.686.

Nordhaus, William. 2017. "Integrated Assessment Models of Climate Change." National Bureau of Economic Research. *The Reporter*. No. 3. September. https://www.nber.org/reporter/2017number3/integrated-assessment-models-climate-change.

Oswald, Andrew, and Nicholas Stern. 2019. "Why Are Economists Letting down the World on Climate Change?" *VoxEU.Org*, September 17, 2019. https://voxeu.org/article/why-are-economists-letting-down-world-climate-change.

Perman, Roger, Yue Ma, James McGilvray, and Michael Common. 2003. *Natural Resource and Environmental Economics*. Pearson. Addison Wesley.

Petrie, Peter and Vinod Thomas. 2013. Development Imperatives for the Asian Century. ADB Economics Working Paper Series. No. 360. July. https://www.adb.org/sites/default/files/publication/30306/ewp-360.pdf.

Picciotto, Robert, and Vinod Thomas. 2021. "Opinion: The Real Problem in the World Bank's Doing Business Indicator." *Devex*, October 19, 2021. https://www.devex.com/news/sponsored/opinion-the-real-problem-in-the-world-bank-s-doing-business-indicator-101848.

Plumer, Brad. 2022. "*U.S. Greenhouse Gas Emissions Bounced Back Sharply in 2021*". New York Times. https://www.nytimes.com/2022/01/10/climate/emissions-pandemic-rebound.html.
Quah, Euston. 2015. "Pursuing Economic Growth in Asia: The Environmental Challenge." *The World Economy* 38 (10): 1487–1504. https://doi.org/10.1111/twec.12352.
Sathirathai, Suthawan. 2001. "Valuing Mangrove Conservation in Southern Thailand." *Contemporary Economic Policy* 19 (2): 109–122. https://doi.org/10.1093/cep/19.2.109.
Searcey, Dionne. 2021. "What Caused the Blackouts in Texas?." *The New York Times*, February 17, 2021. https://www.nytimes.com/2021/02/17/climate/texas-blackouts-disinformation.html.
Stern, Nicholas. 2007 Stern, N. H. 2007. *The Economics of Climate Change: The Stern Review*. Cambridge, UK: Cambridge University Press.
Stern, Nicholas, Joseph E. Stiglitz, and Charlotte Taylor. 2021. "The Economics of Immense Risk, Urgent Action and Radical Change: Towards New Approaches to the Economics of Climate Change." *National Bureau of Economic Research*, NBER Working Paper Series, 28472. https://www.nber.org/system/files/working_papers/w28472/w28472.pdf.
Stern, Nicholas and Joseph Stiglitz with Charlotte Taylor. 2022. "The Economics of Immense Risk, Urgent Action and Radical Change: Towards New Approaches to the Economics of Climate Change." *Journal of Economic Methodology*. https://doi.org/10.1080/1350178X.2022.2040740.
The Economist. 2019. "A Bold New Plan to Tackle Climate Change Ignores Economic Orthodoxy." *The Economist*, February 7, 2019. http://www.economist.com/finance-and-economics/2019/02/07/a-bold-new-plan-to-tackle-climate-change-ignores-economic-orthodoxy.
Thomas, Vinod. 2019. "The Necessity of Climate Economics | by Vinod Thomas." Project Syndicate, October 16, 2019. https://www.project-syndicate.org/onpoint/economic-growth-models-embrace-climate-change-by-vinod-thomas-2019-10.
Thomas, Vinod, and Namrata Chindarkar. 2019. *Economic Evaluation of Sustainable Development*. Singapore: Springer Singapore. https://doi.org/10.1007/978-981-13-6389-4.
Thomas, Vinod, Mansoor Dailimi, Ashok Dhareshwar, Daniel Kaufmann, Kishor Nalin, Ramon Lopez, and Yan Wang. 2000. "The Quality of Growth." *World Bank Publications*. https://doi.org/10.1596/0-1952-1593-1.
United Nations. United Nations. 2007. "2007— Intergovernmental Panel on Climate Change (IPCC) and Albert Arnold (Al) Gore Jr." https://www.un.org/en/about-us/nobel-peace-prize/ipcc-al-gore-2007.

Yale News. 2018. "Yale's William Nordhaus Wins 2018 Nobel Prize in Economic Sciences." *YaleNews*, October 8, 2018. https://news.yale.edu/2018/10/08/yales-william-nordhaus-wins-2018-nobel-prize-economic-sciences.

CHAPTER 8

Integrating Resilience in Policymaking

Running away from a problem only increases the distance from the solution.
Anonymous

Policies for resilience building affect different levels of policymaking from economy-wide macroeconomic measures to sector and project specific microeconomic ones. They have impacts on multiple themes ranging from financial stability to food security. In the throes of the climate crisis, integrating environmental care into development policy involves adaptation and mitigation. While these are complementary strategies, each is necessary for responding to climate change. Without mitigation, global warming will keep escalating until a point is reached where no amount of adaptation will help.

Raising the bar for resilience building starts by going to the underpinnings of rising risks and then moving forward with solutions that exploit synergies across related policy areas to tackle these risks. Resilience building has elements of securing robustness but importantly also flexibility of systems to adapt and respond. In the middle of a disaster, not only the speed of delivery but also care in following protocols are crucial for good outcomes, much as the performance of a medical emergency depends on both the speed of attention to the patient and adherence to procedure. The reconstruction phase dovetails with pre-disaster as systems need to be made functional while at the same time new events will need to be anticipated and planned for.

© The Author(s), under exclusive license to Springer Nature
Singapore Pte Ltd. 2023
V. Thomas, *Risk and Resilience in the Era of Climate Change*,
https://doi.org/10.1007/978-981-19-8621-5_8

Crisis and Disaster Management

The world has faced macroeconomic crises periodically, and lessons have been learnt on how to be prepared for them and to handle them better. The year 2022 is seeing a confluence of macroeconomic challenges, marked by the resurgence of inflation after a period of considerable stability on this front. Both the pandemic and the war have played a big role in aggravating inflationary pressures (Gopinath 2022). Building resilience for the global macroeconomy is a balancing act of keeping price increases in check while at the same time avoiding a global recession.

The task of strengthening resilience cuts across the various themes and sectors of the global economy. The SDGs summarise the major goals of policy, and they demonstrate how disaster risk reduction is linked with well-being and development. Because disasters push people into income and consumption poverty, as well as cause other disruptions to livelihoods and society, achieving the SDGs will be difficult if disaster risk reduction is not prioritised.

From the policy perspective, governments need to urgently recognise the increasing incidence of disasters. The response should be strengthened and better investments made in preventing and reducing disaster risks. For every US$100 of official development assistance, only 40 cents have gone to disaster risk reduction (Thomas 2017). Protecting human and physical capital from disasters should be regarded as a public service worthy of spending, say, 2% of national budgets.

Disaster management cycles, as described by various authors, can be divided into three important phases: pre-disaster, disaster relief, and reconstruction and recovery (Fig. 8.1). Policymakers need to do much more to prepare for disasters rather than reacting only after they strike. It helps to mobilise funds ahead of time so that they can be swiftly disbursed for disaster management. It is important to build capacities across institutions for disaster monitoring and response before a disaster strikes.

Progress is being made in tackling the priorities in disaster risk management in many countries (ADB 2021a). Unfortunately, the resilience bar keeps getting higher as the decades roll by. In the era of climate change, resilience building involves decarbonising the economy while attending to the concerns over economic growth and well-being. Box 8.1 briefly looks at Vietnam's effort to integrate climate resilience into development plans.

Fig. 8.1 Phases of a disaster (*Source* Author's depiction based on the literature)

Box 8.1 Vietnam's Resilient and Decarbonising Pathways

Vietnam has announced its economic goal of achieving high-income status by 2045. Yet, as one of the most vulnerable countries to acute weather, concerns over the havoc that climate change can wreak on Vietnam's population, livelihoods, and infrastructure, pose a serious obstruction to this goal. Together with Bangladesh, Vietnam is ranked at the top when it comes to exposure to flooding. The country is also highly exposed to tropical cyclones.

Vietnam is extremely susceptible to floods and erratic temperatures due to its low-lying land and proximity to large water bodies. It risks losing a significant portion of its annual GDP to damages caused by climate change. The World Bank estimated Vietnam's GDP loss to climate change to be about 3.2% of total GDP in 2020 and expects this to rise to 12–14.5% by 2050, if no adaptation or mitigation measures are taken. This economic damage also means that poverty levels are likely to increase.

Thus, a dual pathway of resilience and decarbonisation has merit in helping Vietnam achieve its economic goals, while simultaneously reducing its vulnerability to the rising probability of damage from climate change. The resilience pathway advocates adaptation measures such as infrastructural buffers, fiscal policy reforms, and social programmes to increase Vietnam's physical and social strength against the effects of climate change.

> The decarbonising pathway encourages policy and financial investments in green practices, such as carbon taxing, and alternative forms of energy. Both measures require significant economic input from domestic savings, and public and private funding. This is necessary short-term spending for longer-term economic returns. So, if Vietnam wants to achieve its economic ambition, climate action and socio-economic investments must go hand in hand.
> *Source* The World Bank (2022).

Infrastructure to the Rescue

Investing in early warning systems and robust evacuation plans has been shown to have high and tangible benefits from saving lives relative to the costs of investing in them. Improved early warning systems and disaster management are credited with reducing the number of deaths across regions from natural disasters almost threefold between 1970 and 2019 (UNDRR 2021). There is an important role for remote sensing technologies in improved warning systems.

Cyclone Bhola, with wind speeds of 200 kilometres per hour, took some 170,000 lives by official estimates (and 500,000 by unofficial ones) in 1970 (Roy 2014). Bangladesh invested US$10 billion roughly between 1980 and 2015 in cyclone readiness, equipping the country with early warning systems, disaster-resilient shelters, and embankment protection (World Bank 2016; World Meteorological Organization 2020). The payoff was remarkable. When Cyclone Sidr hit Bangladesh in 2007, with wind speeds of 240 kilometres per hour, 3406 lives were lost (Government of Bangladesh 2008).

Japan has demonstrated how warning systems can be strengthened to locate vulnerable members of the population and enable their escape to disaster shelters. Japan's Meteorological Agency, in 2013, updated its Emergency Warning System to underscore the imperative for evacuations and to map the intensity of weather-related hazards and people's special needs (Japan Meteorological Agency 2013).

While countries are improving their national emergency capabilities, central governments often defer to their local counterparts on the choice of the systems to adopt. Turkey's National Emergency Management Information System, together with its Uninterrupted and Secure

Communication System Project, link authorities during disasters and emergency situations. Advances in ground-based networks of radars, but increasingly also satellite data, are vital for being able to continuously observe global weather.

Satellites provide information for wide geographic areas, including oceans, improving forecasting and making warning systems more efficient. Communication capabilities for emergency warnings are expanding rapidly because of the explosive development in mobile networks. For example, Australia's Emergency Alert enables states and territories to issue warnings to landline and mobile phones linked to buildings in high-risk areas, and it works across all telecommunication carrier networks. Technologies that link sensor networks, large-scale data analysis, and communications systems can provide decision makers with timely information to guide responses.

Siemens installed a levee monitoring system in the Netherlands using sensors to monitor water pressure, temperature, and shifting weather patterns. It helps to identify thresholds that are at risk of being breached and trigger alarms. IBM has developed a digital command centre in Tamil Nadu, India that integrates real-time information on storm conditions, emergency response assets, and areas at risk.

A good example of the value of preparedness through robust evacuation plans emerged during the experience with Typhoon Haiyan, arguably the biggest typhoon on record to make landfall. The entire population of Tulang Diyot—a tiny island off mainland Cebu in the Philippines—was saved from the typhoon by enforced pre-emptive evacuation. Although no houses were left standing, all 1000 inhabitants were rescued (UNDRR 2013). Similar stories emerged from the 2004 tsunami in Indonesia's Aceh region, where multipurpose, tsunami evacuation centres were built since then (Yuzal et al. 2017).

Strengthening the capacity of people, lifelines, and infrastructure to withstand and rapidly recover from hazards will limit losses and disruptions and can even prevent hazards from turning into disasters. Aside from early warning systems, resilient and accessible infrastructure for safe water, hospitals, and evacuation centres are among the most important investments. Breaks in these lifelines are the major causes of desperation and breakdowns in law and order that often follow climate-related disasters and earthquakes.

Facilities vital to crisis response must be linked to networks that will not fail populations. Seismic retrofitting of hospitals in Sendai, Japan, enabled people to provide health services after the 2011 Tohoku earthquake and tsunami. Disaster-proofing hospitals, by one measure, adds less than a tenth to the cost of new hospitals, and eliminates the need to rebuild that would virtually be double the initial cost. Raising the base of houses above flood levels—by putting them on pillars or using higher foundations—can enhance the resilience of residents and structures.

In Bangladesh, a post-flood housing reconstruction project introduced capping traditional earth plinths with cement-stabilized soil. This proved highly effective in subsequent floods. The Philippine Geosciences and Mines Bureau's geohazard mapping programme identifies communities at risk from the landslides and flash floods that are so often triggered by seasonal typhoons and storms.

Local business entities can be enormously helpful in upgrading early warning capabilities. Companies can help to make infrastructure more resilient, improving the speed of recovery after disasters strike. Backup systems for critical infrastructure and waterproof or diesel-powered pumping systems reduce the chance of water and power system failures. System intelligence is another form of building resilience. Embedding sensors and controls into power lines and water treatment plants can allow cities to assess hazardous conditions, take preventative measures, and target repair efforts.

Because no country can have enough people and financial resources in the exact place where a disaster strikes, better coordination is needed to get resources in place quickly. Since the 2011 earthquake and tsunami, Japan has been improving the ties between the national government that oversees policy and local governments in charge of implementation (Ishiwatari 2021). The use of logistical and technological capabilities of the private sector is also a big part of such preparedness.

Society and Governance

Changing mindsets will be essential for disaster prevention to be treated as a policy priority. Not just that, governments must be prepared to spend adequately on crisis preparedness. Unfortunately, the political economy does not seem to favour attention to preparing for a disaster compared to rescue and restoration. One aspect of this dilemma is that voters seem to recognise the delivery of relief spending as opposed to spending in preparedness (Healy and Malhotra 2009).

Cities are central to achieving a more sustainable development (Yusuf 2011). High density of populations makes cities the primary sources of emissions. By the same token, breakthroughs in sustainable development can come from innovations in urban managements. Greater energy efficiency, better waste management, and management of pollution and congestion can be central to smart city management. Innovations in disaster management are also part of good urban governance.

Regardless of the nature of the disaster risk, interventions that bolster health and education strengthen preparedness. Both areas rely on the trust and participation of the community in their institutions and policy (Maya 2016). Singapore and South Korea, two countries that have had strong economic growth, also stand out for the size and quality of health and education spending (public and private) that make lasting contributions to building resilience. Payoffs to investing in education, information sharing, and capacity development have been high in dealing with disasters.

Education, awareness-raising, and capacity development will be vital to all phases of disaster risk reduction. These interventions improve the quality of responses, strengthen compliance to directives, and advance innovation in risk management. Relief and recovery have features of an emergency, but they too need sound protocols and modern technologies. In reconstruction, mitigation should be integral to infrastructure investments. Governments may be the lead actor in resilience building, but businesses, households, and all segments of society need to be engaged for effective and sustainable results.

With the increased frequency of floods and storms, the availability of financial resources is a vital part of disaster management. The Philippines' catastrophe risk insurance programme buffers central and local government agencies against the financial losses from severe natural disasters. Providing US$206 million in aggregate coverage, the programme protects assets of the national government and 25 provinces. Under the programme, a government-owned insurance agency provides catastrophe risk insurance to the national government and the participating provinces (World Bank 2017a).

Resources also need to be set aside for regulatory policy. Zoning regulations are vital to disaster readiness. With rising sea levels and temperatures, previous norms of the safe distance to live from the coastline must be revised. A part of this effort would be to restrict new development in hazard-prone areas and building codes to protect businesses, homes, and neighbourhoods. With changing circumstances, local decision makers need effective and innovative land use planning to manage risks.

It has also become essential to minimise the kind of disruption to supply chains and information networks seen during the massive floods in Sri Lanka, India, and Thailand in the 2010s. Global supply chains involve producers, suppliers, and distribution links, which are naturally vulnerable to disasters. (UNESCAP-UNISDR 2012). With globalisation, the protection of supply chains is a major priority in business plans. Since the outbreak of COVID-19, major supply chain disruptions have been seen across the world. For example, the pandemic has hurt supply conduits for vaccines, protective equipment, and food, calling for steps to strengthen the resilience of supply chains and trade logistics (Park 2022).

As resources are channelled to crisis management and resilience building, impact evaluation, cost–benefit analysis, and objectives-based evaluation are important tools for governments to understand the effectiveness of their development investments (Thomas and Chindarkar 2019). This is especially so when resources are scarce and need to be judiciously used to achieve sufficient development impacts and to be able to draw policy lessons. Aside from making obvious economic sense, benefits also accrue to social inclusion and environmental governance.

Indeed, the importance of inclusion and governance continues to be overlooked. Governments and external financiers need to facilitate credit for rebuilding lives and livelihoods, especially for the poor and vulnerable. Good governance and stable institutions are enabled by applying evaluations that connect the dots between programmes and their impacts to be able to understand better if and how their goals are being achieved. The World Bank, for instance, has partnered with governments on resilience building initiatives and to keep an eye on the results of these projects, with the intention of sustaining their benefits (Box 8.2).

Box 8.2 Mozambique Sustainable Irrigation Development Project

More than 75% of Mozambique's population's main source of income is derived from agriculture. Dominating this population are smallholder farmers at 95%. However, despite the significant population reliance on agriculture, only 10% of Mozambique's total arable land is being used, leading to limited productivity.

In addition, Mozambique is highly vulnerable to climate-related risks, such as floods, cyclones, and droughts. With climate change, this is expected to worsen. This means the impact on agricultural households is significant, given the stresses on irrigation in times of drought. This is not to say that Mozambique lacks water resources—on the contrary, but insufficient is irrigation infrastructure. Improving this aspect will improve farmers' agricultural survivability during droughts, as well as increase overall farming productivity and thus returns to livelihood.

In partnership with the World Bank, the Mozambique Sustainable Irrigation Development Project (PROIRRI) was approved in 2011. It particularly supported smallholder farmers to improve the efficiency of water usage, diversify farming systems to reduce vulnerability to climate-related changes, and yield-planning in accordance with market demand. Technical assistance was provided by the World Bank, and impact evaluation was provided with a budget of US$1.35 million.

Iterative process evaluation and feedback on the resilience building process has helped farmers improve their water management methods. PROIRRI is consistently refined and improved based on pilots on water irrigation methods in communities, helping Mozambique gather information on good practices that can be applied to the wider community for more productive use of its arable land, and guard against weather-related disasters.

Source The World Bank (2017b).

DISASTER PREVENTION AND MITIGATION

A lesson of wide relevance is that it is essential to put in place not only short-term responses to save lives but also longer-term investments to rebuild livelihoods. The needed attention to the immediate ought not to crowd out what is important for the long term. The priority must be to save lives. But when things are rushed and resources are tight, it is nevertheless valuable to ensure that immediate steps feed into long-term solutions.

The benefits of slowing global warming, such as averting water-shortage crises, might be fully visible only over time, but interventions are needed now to avoid costly corrections. There is great value in using a part of post-crisis spending for investments—in transport, logistics and electricity grids, in basic healthcare, in skilling and reskilling individuals, and in green technologies and digitalisation.

In turbulent times after crises, careful counterfactual analysis is more important than ever for measuring the impact of disaster interventions. The direct and indirect effects of these interventions must be assessed both for the social well-being and economic performance they bring to beneficiaries. Country programme outcomes are more often conditioned by interventions outside projects, some within and some beyond the control of many players.

Gains are impressive when vital links among related areas are capitalised on. The rebuilding of lives and livelihoods post-pandemic spans multiple sectors and themes. During the pandemic, getting vaccines to people requires not only well-staffed health clinics but also efficiently maintained cold chain logistics, as well as public awareness of the processes. Synergies among multiple sectors must be exploited, especially public–private partnerships, to improve service delivery. Regional and local approaches may offer lessons for strengthening the aid architecture in multi-country efforts.

There are many examples of successful mitigation, but most proposals to take them to the next level are progressing slowly. Multilateral development banks, including ADB, the World Bank, and the IMF, have a huge role to play in enabling low-cost financing and knowledge resources. These institutions have announced substantial increases in climate lending. The implementation and effectiveness of MDB-supported projects need to be monitored, while going much further in the lending plans.

Mitigation and Adaptation Efforts

Figure 8.2 makes the point that incurring the costs of both adaptation and mitigation contributes to lowering the cost of climate disasters. The cost of their impacts is shown to be dependent on the cost of adaptation and the cost of mitigation that are sustained. There is a degree of complementary between these two aspects of climate policy in lessening the damages. At the same time expenditures on mitigation and on adaptation compete within a resource envelope. The extreme case of only accepting the cost of adaptation leads to the highest cost of impacts within this framework.

Adaptation and mitigation responses are underpinned by common enabling factors. These include effective institutions and governance, innovation and investments in environmentally sound technologies and infrastructure, sustainable livelihoods, and behavioural and lifestyle choices. Innovation and investments in environmentally sound infrastructure and technologies can reduce GHGs and enhance resilience to

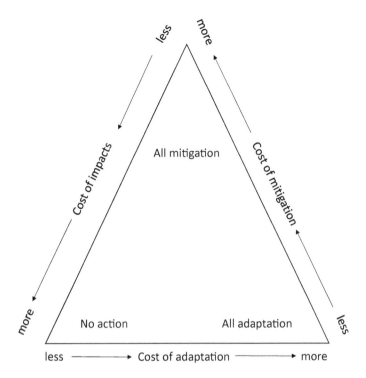

Fig. 8.2 Mitigation and adaptation (*Source* Taken from Parry [2009])

climate change. Innovation and change can expand the availability and effectiveness of adaptation and mitigation options.

Shifts towards more energy-intensive lifestyles can contribute to higher consumption of energy and resource, driving greater energy production and GHG emissions, and increasing mitigation costs. Emissions can, however, be substantially lowered through changes in consumption patterns. Improving institutions as well as enhancing coordination and cooperation in governance can help overcome regional constraints associated with mitigation, adaptation, and disaster risk reduction.

What Affects Success?

Several factors influence the effectiveness of mitigation and adaptation. One is population growth and urbanisation. Viewed from the needs of adaptation, the concern is over increases in the exposure of human populations to climatic variability and change as well as demand for, and pressures on, natural resources and ecosystem services. For mitigation, the question is how this drives economic growth, energy demand, and energy consumption, that result in increases in GHG discharges. Climate responses are also affected by knowledge, education, and the adequacy of human capital.

Gaps in institutions and communication lessen societal and individual perceptions of climate risks and thus the perceived costs and benefits of different options. These gaps also affect the willingness to change behaviour and reduce emissions. The availability and effectiveness of climate finance also affect progress on adaptation and mitigation. Such finance covers local, national, and transnational sources, both public and private, that are directed at mitigation or adaptation or both. Lack of financing clearly reduces the scale and timeliness of investment in climate policies; it also has a bearing, particularly for developing economies, on their capacity to cut emissions.

Inequalities are an important determinant of climate risks as the burden of the damages falls disproportionately on poorer countries and on poorer segments of the population (Guivarch et al. 2021). As mentioned in Chapters 3 and 4, income inequality and poverty limit the ability of the most vulnerable to adapt and cope when disasters strike. At a global level, inequalities constrain the ability of developing economies—especially those with low incomes and a high incidence of poverty—to be able to take vitally needed steps in adaptation and to make a meaningful contribution to GHG mitigation.

Economic Policies for Risk Management

In thinking of ways to address the societal damages of economic activity, one approach is to stress institutional responses which have the capacity to influence society's behaviour and tackle the root causes of damage. Voluntary agreements, underpinned by education and communication campaigns, can take many forms, and, in principle, make a difference.

Institutional arrangements involve the engagement of stakeholders and meaningful negotiations but doing this carries the risk of delays and ineffectiveness, which need to be forthrightly addressed. Bargaining is nevertheless an option especially when fewer affected parties are involved. If bargaining does offer substantial efficiency gains, governments can facilitate achieving these gains by defining and allocating property rights where that is practicable. Governments should provide an institutional structure that promotes bargaining, but this presents its own challenges. Bargaining needs to be between current and representatives of future generations as behaviour today imposes costs on future generations.

Environmental goals can be furthered by encouraging people to behave responsibly. The idea that individuals do not exclusively act in a narrowly utilitarian way suggests this objective may have promise. The success of ethical investment funds, and people's willingness to support charities, are cases in point. Social responsibility can also be underpinned by a variety of approaches. An example is environmental labelling, with volatile organic compound emissions in Germany (Global Regulation 2004), and paper recycling in South Korea (US EPA 2001).

A second approach could be to emphasise regulations and restrictions, for example, quantitative restrictions or controls more akin to command-and-control policies. Examples include input controls over the quantity or mix of inputs; emission standards and licences; output quotas or prohibition and location controls, such as zoning; planning controls; and relocation. The dominant method of reducing pollution in most countries is the use of direct controls over polluters. This can be stipulated as the maximum allowable discharges or the minimum amount that must be cut.

Third, there are merits in using market mechanisms of prices, taxes, and subsidies. The dual objective of climate mitigation and spurring economic growth can be achieved by imposing a price on CO_2 emissions that reflects the burden these emissions impose on the environment. Carbon pricing is a way, rightly favoured by economists for their efficiency, to discourage damaging carbon discharges (Stiglitz and Stern 2017). Under carbon pricing, one approach is using emissions trading schemes, giving policymakers control over consequent emissions levels. There are challenges in designing the specifics of carbon trading. That said, it can encourage polluters to consider alternative, more sustainable energy sources, since the price of pollution has now increased (Black et al. 2022).

A tax on companies or households that pollute, is another way to price carbon. It has long been the preferred means advocated by economists to achieve a pollution target (Jorgenson and Wilcoxen 1995), noting that it has limits and is only a part of the package of needed measures (Green

2021). The crucial questions would be the choice of the tax rate, effectiveness of implementation and the subsequent use of the revenues. A question often asked is whether carbon taxes are high enough to make a difference to the discharge of emissions from fossil fuels or if subsidies are sufficient to encourage non-polluting energy sources, such as renewables.

One way to do this is to add a carbon tax to the price of polluting products like petroleum or steel. Ideally the tax would be equivalent to the estimated damage associated with the discharge corresponding to a unit of production. If the tax is big enough, producers would cut pollution to avoid paying the tax. The highest rate has been set by the Danish government for 2025 increasing to €150/tCO_2 in 2030 (Enerdata 2022). But the global average is only US$6 a tonne of CO_2. Singapore is emphasising carbon taxation as part of its climate policy (Box 8.3). High and effective carbon taxation across major polluters like China, the US, India, Russia, Japan and Germany (which together account for 60% of global effluents) could make a big difference to global pollution (Rapier 2019).

To the extent that some pollution continues with the tax being paid, the tax revenue raised could be used to support cleaner fuels. So, carbon taxes can not only discourage high-carbon production, but also raise revenue for green growth. A higher price of carbon will boost incentives for investment in low-carbon or carbon-reducing technologies, creating longer-term environment-friendly economic operations. Governments can also finance much needed adaptation such as coastal protection or disaster risk reduction more generally. A part of the revenues can also be used to finance safety nets for the poor who may be hurt by short-term rise in energy costs.

Box 8.3 Singapore and Carbon Tax

Singapore accounts for only 0.1% of the global carbon footprint, but it has high emissions per person—the 27th highest out of 142 countries, as of 2018. Singapore is reliant on natural gas and has geographical limitations in switching to solar and wind (Mohan 2021). This is even more reason to scale up carbon tax to encourage energy producers to cut emissions.

The effect of a tax on effluents depends both on the tax rate and the extent of its coverage of the pollution sources. In 2019, Singapore set the carbon tax at S$5 per tonne. This was at the low end of a range that

> spans Japan's US$2.60 (S$3.50) per tonne to Sweden's US$137 (S$184) per tonne. On the other hand, Singapore's tax covers four-fifths of the country's emissions compared with only one-third in the EU.
>
> A solid case can be made for Singapore to sharply raise its carbon tax to S$50 per tonne. In 2020 the government announced S$25 for 2024, and S$45 by 2026. It could possibly rise to S$80 by 2030. On the tax revenues, there would be a decarbonisation fund as well as plans for a carbon dividend that would benefit the consumers. Complementing the tax should be investments to make renewable energy, such as solar and wind, far more viable. Singapore plans to quadruple solar deployment by 2025, although this comes from an extremely low 2020 base.
>
> *Source* Thomas (2022a).

In talking about effluents, the discussion is usually about those that are related to production within a country's borders. But roughly one fifth of the global emissions can be identified as being import-related. Thus, when considering emissions linked to consumption rather than just domestic production, a carbon tax should also target discharges contained in imports that are usually excluded from country contributions. Doing so makes a difference. The divergence between consumption and production-based emissions has been rising. For example, instead of a 3% increase in production-based emissions since 1990, the US would have a 14% increase if the measurement is consumption-based (Ritchie 2019).

The question might be asked if a full-fledged application of pricing and market tools could prevent the worst outcomes. The interventions considered here would be ones that are supported by textbooks in neoclassical economics. Imagine if:

- Countries stop using the faulty, gross measure of economic growth, **GDP**, and at least complement it by a measure that nets out damages from externalities.
- All countries adopt **carbon pricing**, for example, via a significant carbon tax that is levied on the source of the pollution.
- All development **projects** pass a climate test and are required to be accompanied by legal covenants on mitigation and adaptation.

- A high enough **quantitative restriction** is placed on fossil fuels, in addition to eliminating all subsidies for this pollution source, while subsidies go to clean energy in accordance with its social benefits.
- Given the constraints on transition and financial intermediation, high-income countries provide vast **climate financing** to low-income countries, facilitated by an unprecedented alliance among MDBs, especially the IMF, the World Bank, Asian Development Bank, and New Development Bank which have strong climate mandates.

Such a package should have a decisive impact on decarbonisation in the near term. Despite the analytical case for such an agenda, the roadblocks are immediately seen as being political, with the minority of the losers from the policy reform lobbying to discredit and block the change. That in turn is linked to people's preferences, perceptions, and valuation of the individual versus the society, and the present versus the future. That is why the closing chapter returns to the theme of what it might take to see transformational change.

Alternative Energy Sources

Global energy-related CO_2 emissions continue to rise, for example, by over 2 billion tonnes in 2021, driven by a strong rebound in demand for coal in electricity generation (IEA 2021a). Since energy production and use are the largest sources of GHGs, energy is central to achieving climate goals, which in turn are critical for disaster risk reduction. Exploration of new technologies, such as green hydrogen or modern versions of safer nuclear energy, not only in their development but also commercialisation, are priorities.

Greater energy efficiency is one of the tools for lowering global energy consumption and slowing global warming and extreme disasters. Progress has been made in energy efficiency since 2005, at least until 2015, when this regressed. Improvements since 2018 have been well below what is needed to achieve the SDGs' sustainability goals. Energy efficiency is envisaged to make up more than 40% of the cut in energy-related GHGs over the next 20 years in one scenario (Motherway 2020).

Renewables offer immense potential for enabling a low-carbon transition. The costs of generating electricity from renewables have fallen sharply. The speed at which renewable plants are built these days is an advantage. Wind farms take 9 months to put up and solar parks can be built in just 3–6 months. Coal, gas, and nuclear plants take years to build. Countries that need to build additional generating capacity quickly should consider renewable energy. In 2017, 18% of the electricity produced in the US came from renewable sources (Morris 2018).

Nearly 60% of renewable energy were from modern sources (biomass, geothermal, solar, hydro, wind, and biofuels) and the remainder from traditional biomass that is used in residential heating and cooking in developing countries. Renewables made up 28% of world electricity generation in 2021; this is expected to rise substantially by 2040 (IEA 2021b). The development and use of renewable energy technologies will depend heavily on government policies and financial support to make renewable energy cost competitive.

Valuable as they are, the specific energy projects, such as solar and wind energy, are not proving to be nearly enough in time to enable the low-carbon transition in energy. Extreme catastrophes can be averted only if the trajectory of global warming is matched by systemic and radical changes in energy use, and not just the incremental and specific adjustments being undertaken today. This is a tall order, but the alternative is an order of magnitude worse.

Costs and Benefits

Faced with the climate conundrum, vast resources should be committed for low-carbon activities, even—if need be—at the expense of less critical objectives. G-20 economies initially committed US$5 trillion in 2020 for the COVID-19 stimulus packages (Segal 2020). Like the number mentioned in Chapter 1, an estimated figure for spending by the G-20 nations for 2020–2021 is on the order of US$14 trillion (Nahm, Miller, and Urpelainen 2022). Compared with these numbers, the annual capital costs of the needed low-carbon transition would seem doable if there is the political will. Countries also look to multilateral lenders, such as the World Bank and IMF, for leadership in mobilising global climate funds. At the same time, countries need to spend higher proportions of their GDP on disaster management to mitigate internal vulnerabilities and increase resilience (Thomas 2022b).

The IPCC indicated in 2019 that keeping the rise in temperatures below 1.5 °C from pre-industrial levels would involve costs in energy alone of 2.4 trillion (in US$2010) a year (from 2016) through 2035

(Fogarty 2018; IPCC 2019; Yeo 2019). Another estimate indicates that US$6.3 trillion will be needed globally each year until 2030 to build infrastructure in energy, transport, water, and telecommunications (OECD 2022; Bloomberg 2018). Set against the government stimulus in 2020–2021 to tackle the economic impact of COVID-19, this amount of capital investment looks attainable.

To motivate such a shift in priorities, comparisons of the costs of the carbon transition with the associated benefits would help even if they are difficult to estimate for common time frames. But rough orders of magnitude suggest that the economic gains from mitigation far outstrip the financial cost of interventions to strengthen mitigation. By one estimate, averted global indemnities by containing warming to 1.5 °C could be US$150 trillion to US$792 trillion by 2100 (E360 Digest 2020; Kahn et al. 2019; Wei et al. 2020). In the same time frame, one set of projections suggests that global real GDP per capita would fall by 7.2% by 2100 with a continued rise in world temperature by 0.04 °C per year in contrast to a loss of 1.07% with a temperature increase of 0.01 °C under the Paris Agreement (Kahn et al. 2019). Health benefits from slowing global warming in one set of estimates could exceed the investment costs of achieving such mitigation by 1.45–2.45 times (Watts et al. 2019).

The economic benefits of adaptation are known to be large, especially for lower-income groups. For example, by one estimate, investing in resilient infrastructure in developing countries could bring benefits of the order of US$4.2 trillion (Hallegatte et al. 2019). Another assessment is that adaptation investments of US$1.8 trillion globally during 2020–2030 could generate US$7.1 trillion in net benefits (Global Commission on Adaptation 2019). The need for these adjustments, and the cost of putting them in place rise with the passage of time.

The economic slump following the breakout of the pandemic in 2020 brought sharp drops in toxic chemicals in the air. This was seen as a silver lining in the otherwise dark clouds of health catastrophe, and a glimpse of what low-carbon economies could look like. But carbon dioxide, the chief culprit in global warming, rose to a record level in April 2020, and continued to rise thereafter (World Meteorological Organization 2021; Bhanumati et al. 2022). The tipping point has already been reached in some respects in the struggle to keep temperatures from rising 1.5 °C above pre-industrial levels (Janunta 2021).

A common threat to achieving the Paris[1] Agreement targets for mitigation in Asia is the prospect that coal will become the region's largest energy source by 2040. Failure to move away from fossil fuels, especially coal, could very well lead to the failure to meet Paris Agreement commitments on cutting emissions. The way forward is to drastically accelerate the switch to wind, solar and other renewables in national energy mixes, to cut costly fossil fuel subsidies, as India and Indonesia have started to do, and to shun any form of financial links to coal. The urgency for achieving a transition to a low-carbon growth path cannot be overstated. It is already late in the day to be making this transition since accumulated emissions in the atmosphere are already dictating the trajectory of rising temperatures for the coming decades.

GREEN FINANCING

Climate-related risks translate to financial risks in at least three ways. First, they do so through the manifestation of physical risks, such as the increased frequency of severe weather events that may damage property and infrastructure and disrupt trade and economic activity. Second, gradual temperature changes could affect the value of assets. Third, for the banking sector, the impacts may be felt directly through the exposure of mortgage banks to flood risk, or for globally active banks, through the impact of natural disasters on sovereign bond ratings and country risk.

Risks and Rewards

Investing in resilience against rising risks often requires proactive policy interventions. Figure 8.3, following well-known discussions in the literature, shows the channels for intervention both in lowering the risk of the needed investments and in increasing their returns. An example would be a shift from a commercially unattractive investment opportunity (low and right) to a commercially attractive one (high and left). This is achieved in two steps: first, reducing the risk of the activity (for example through a regulatory policy, such as guaranteed access to the grid for independent power producers) and second, by increasing returns on investment (for example, by creating financial incentives, such as premium prices for renewable electricity through feed-in tariffs).

Disaster risk finance is a key pillar of disaster risk management. It cuts across different areas of disaster risks and covers the different

Fig. 8.3 Shifting the risk-reward profile or a renewable project (*Source* Based on Glemarec et al. [2012])

phases of a disaster. The distinct phases also require different types of financing instruments that are timely and cost-efficient. Governments need to have a national financing strategy that anticipates the sources and uses of financing. The financing also concerns different interventions of the national government level and line agencies and local governments (Financial Protection Forum 2018).

The viability of projects for disaster risk reduction is augmented ironically by the spike in disaster damages. In 2021 for instance, around 850 natural-loss events occurred in 2018, including floods, tropical cyclones, wildfires, and earthquakes in the US and Japan, incurring a total cost of US$160 billion, according to Munich RE's NatCatService (2022). There are financial risks for parties that have suffered losses from the effects of climate change and seek compensation from those they hold responsible. Weather-related insurance losses have increased almost five-fold to an average of around US$50 billion per annum so far this decade from an average of around US$10 billion per annum in the 1980s.

In global insurance, gaps in protection offered remain sizeable. The uncertainty associated with climatic scenario analysis complicates the challenge of modelling implications for insurers' liabilities. This affects households and businesses because they could face more expensive or more curtailed insurance policies. There are transition risks as households, businesses and industry sectors face costs, valuation losses, and disruptions from the adjustment to a low-carbon economy. These risks are longer-term and less visible and have yet to materialize. As such, they may not carry a great sense of urgency.

Sources of Climate Funds

The importance of having an adequate level of available financing for climate investment cannot be overstated (Bhattacharya and Stern 2021). This finance can be in the form of loans or grants that draw on the resources of the public and private sectors to support climate interventions. A starting point for official flows is the UN COP16 Accord which states that developed countries commit to a goal of mobilising jointly US$100 billion per year by 2020 to address the needs of developing countries.

The language of climate accords starting with COP16 makes it clear that this amount may include finance from public and private sources. Climate finance has been rising, but it is still short of the annual target in 2020. Starting with attaining the US$100 billion a year by 2020, the climate finance system must scale up, urgently. Sachs (2021) proposed that rich countries be taxed at a rising rate, and differentially between high- and middle-income ones, to finance mitigation and adaptation in poorer countries. One of the expectations of COP27 is that developed countries (public and private sectors) fulfil their commitment to developing countries regarding climate financing pledges, that have thus far fallen short.

Given its global nature and the presence of cross-border spillovers, climate change should be a top financing priority for the MDBs. Their overall mandate should clearly show the priority for tackling the problem of global public goods. This is a stated priority of ADB's current long-term strategy, Strategy 2030. ADB signalled commitments of US$80 billion in climate finance from 2019 to 2030, subsequently raised to US$100 billion, and wants at least 75% of its projects to include mitigation and adaptation components by 2030 (Asian Development Bank 2021a). Despite the strain on resources for investment in the face of the COVID-19 pandemic in 2020, ADB recorded US$4.8 billion in climate finance in 2021, of which roughly 75% was for mitigation and 25% for adaptation (Asian Development Bank 2021b). The World Bank's Climate Change Action Plan 2021–2025 envisages a resilience ratings system to measure and disclose the extent to which adaptation and resilience considerations have been integrated into World Bank–supported project designs.

Along with climate-related lending, MDBs such as the World Bank Group ought to have zero tolerance for financing any fossil fuels, including downstream investments. While international organisations are beginning to take this direction, the IEA's Stated Policies Scenario warns that GHG emissions could continue rising if current policy settings are maintained. This is the cold reality against which mitigation and adaptation projects need to be vastly scaled up in country investment portfolios. Development projects must also be put to test for climate proofing against risks and disasters. All investment projects, domestic and externally funded, will need to pass through the filter of climate proofing. To ensure development benefits, projects need to be robust throughout their lifetimes, taking account of likely impacts.

The potential role of the private sector in mitigation and adaptation is decisive, noting that investment risks in these respects need to be overcome (IMF 2022). Private actors rely primarily on their own balance sheets to finance renewable energy projects. The reasons for investors' reliance on balance sheets vary, including the size of the project (it can make more sense to finance small projects internally), difficulties in securing debt, high costs of capital, and other factors. They can finance climate investment projects by using either on balance sheet financing or borrowing funds from a bank in the form of a loan, or through equity capital from selling a stake in the business itself. Equity investors in renewable energy take equity positions in companies, projects, or a portfolio of projects, and expect a greater return for the level of risk they take. Equity investments are usually in the form of funds and involves many actors. Banks focus on getting debt repaid, earning a relatively small return on transactions. Usually, commercial debt is the cheapest source of finance available to project proponents.

Public actors delivered more than half of their financing in the form of grants and low-cost loans. Public concessional or lower-than-market-rate finance, including loans with longer tenors and grace periods, play a catalytic role by supporting the establishment of better policy frameworks, strengthening technical capacity, lowering investment costs, and reducing risks for the first movers in a market. Over the past decade, the global central banking community has been busy repairing and reforming the financial system after the global financial crisis of 2007–2008. Now, it is turning its attention increasingly to the daunting challenge of improving climate resilience as it relates to the financial system and the economy (OMFIF 2019).

Conclusions

A key aim of this chapter has been to draw attention to the policy priority of building disaster resilience, especially spending more and better ahead of disasters, investing in disaster risk reduction, and building capacity for relief and recovery. With the increased frequency of weather disasters, the availability of financial resources will be a deciding factor. Investing in early warning systems and robust evacuation plans, with proven payoffs in benefits relative to costs, should become a centrepiece of the agenda. The success of resource mobilisation may depend on identifying sources of funds, analysing investments for disaster risk reduction, and building capacities for monitoring and evaluation.

It is essential to put in place not only short-term responses to save lives but also longer-term investments to rebuild livelihoods and put development on a sustainable path. Policymaking must recognise the central role of harmful activities that characterise the growth process. One way is to integrate the costs and benefits of environmental stewardship and climate action. This is not about slowing economic growth but ensuring a better quality of growth that is sustainable. That involves avoiding negative spillovers from translating into more frequent setbacks, and countries fortifying the capacity of their systems to handle reversals on more than one front.

Of great importance are effective institutions and governance, innovation and investments in environmentally sound technologies and infrastructure, sustainable livelihoods, and behavioural and lifestyle choices. Better coordination is needed to make the best use of resources, including getting them to the right places at the right time. Strengthening the capacity of people, lifelines, and infrastructure to withstand and rapidly recover from a hazard will limit losses and prevent hazards from turning into disasters. Changing frames of mind, awareness-raising, and capacity development will be key to all aspects of disaster risk reduction.

Note

1. The Paris Agreement on climate change sought to hold the increase in the global average temperature to well below 2 °C above pre-industrial levels, if not below an increase to 1.5 °C.

Bibliography

Asian Development Bank. 2021a. "Six Ways Southeast Asia Strengthened Disaster Risk Management." *Asian Development Bank*, May 4. https://www.adb.org/news/features/six-ways-southeast-asia-strengthened-disaster-risk-management.

Asian Development Bank. 2021b. "ADB's Focus on Climate Change and Disaster Risk Management: Strategy and Programs." *Asian Development Bank*, November 4. https://www.adb.org/what-we-do/themes/climate-change-disaster-risk-management/strategy.

Bhanumati, P., Mark de Haan, James William Tebrake. 2022. "Greenhouse Emissions Rise to Record, Erasing Drop during Pandemic". *IMF Blogs*, June 30. https://www.imf.org/en/Blogs/Articles/2022/06/30/greenhouse-emissions-rise-to-record-erasing-drop-during-pandemic.

Bhattacharya, Amar, and Nicholas Stern. 2021. "Our Last, Best Chance on Climate." *Finance & Development*, September. https://www.imf.org/Publications/fandd/issues/2021/09/bhattacharya-stern-COP26-climate-issue.

Black, Simon, Ian Parry, and Karlygash Zhunussova. 2022. "More Countries Are Pricing Carbon, but Emissions Are Still Too Cheap." *IMF Blog*, July 21. https://blogs.imf.org/2022/07/21/more-countries-are-pricing-carbon-but-emissions-are-still-too-cheap/.

Bloomberg. 2018. "Scientists Call for US$2.4 Trillion Shift from Coal to Renewables, Energy & Commodities." *The Business Times - Energy & Commodities*, October 8. https://www.businesstimes.com.sg/energy-commodities/scientists-call-for-us24-trillion-shift-from-coal-to-renewables?amp.

E360 Digest. 2020. "Countries Must Act on Climate or Risk Up to $792 Trillion in Economic Damage." *Yale Environment 360*. https://e360.yale.edu/digest/countries-must-act-on-climate-or-risk-up-to-792-trillion-in-economic-damage.

Enerdata. 2022. 27 June. https://www.enerdata.net/publications/daily-energy-news/denmark-will-introduce-corporate-carbon-tax-2025.html.

Financial Protection Forum. 2018. "Disaster Risk Finance: A Primer:Core Principles and Operational Framework." August 16. https://www.financialprotectionforum.org/publication/disaster-risk-finance-a-primercore-principles-and-operational-framework.

Fogarty, David. 2018. "Limiting Global Warming to 1.5 °C Possible but Will Need Unprecedented Societal Changes: UN Panel | The Straits Times." *Channel News Asia*, October 9. https://www.straitstimes.com/singapore/limiting-global-warming-to-15-deg-c-possible-but-will-need-unprecedented-societal-changes.

Glemarec, Yannick, Wilson Rickerson, and Oliver Waissbein. 2012. "Transforming On-Grid Renewable Energy Markets: A Review of UNDP-GEF

Support for Feed-in Tariffs and Related Price and Market-Access Instruments." *UNDP-GEF.* https://www.osti.gov/biblio/22090461.
Global Commission on Adaptation. 2019. "Adapt Now: A Global Call for Leadership on Climate Resilience." Rotterdam and Washington, DC: Global Center on Adaptation and World Resources Institute. https://gca.org/global-commission-on-adaptation/report.
Global Regulation. 2004. "Machine Translation of Chemical regulation on the limitation of emissions of volatile organic compounds (VOC) by restricting the placing on the market of solvent-based paints and varnishes' (Germany)." https://www.global-regulation.com/translation/germany/385986/chemical-regulation-on-the-limitation-of-emissions-of-volatile-organic-compounds-%2528voc%2529-by-restricting-the-placing-on-the-market-of-solvent-based-p.html.
Gopinath, Gita. 2022. "How Will the Pandemic and War Shape Future Monetary Policy?" *Presented at the Jackson Hole Symposium*, August 26. https://www.imf.org/en/News/Articles/2022/08/26/sp-gita-gopinath-remarks-at-the-jackson-hole-symposium.
Government of Bangladesh. 2008. "Cyclone Sidr in Bangladesh: Damage, Loss, and Needs Assessment for Disaster Recovery and Reconstruction - Bangladesh." *ReliefWeb*, April 30. https://reliefweb.int/report/bangladesh/cyclone-sidr-bangladesh-damage-loss-and-needs-assessment-disaster-recovery-and.
Green, Jessica F. 2021. Does Carbon Pricing Reduce Emissions? A Review of Ex-Post Analyses. *Environmental Research Letters* 16 (4): 043004. https://doi.org/10.1088/1748-9326/abdae9.
Guivarch, C., N. Taconet, A. Mejean. 2021. "Linking Climate and Inequality". Finance and Development. *IMF*, September. https://www.imf.org/en/Publications/fandd/issues/2021/09/climate-change-and-inequality-guivarch-mejean-taconet.
Hallegatte, Stephane, Jun Rentschler, and Julie Rozenberg. 2019. "Lifelines: The Resilient Infrastructure Opportunity". In *Sustainable Infrastructure Series*, Washington, DC: The World Bank. https://doi.org/10.1596/978-1-4648-1430-3.
Healy, Andrew, and Malhorta, Neil. 2009. "Myopic Voters and Natural Disaster Policy." *American Political Science Review* 103 (2009): 387–406. https://digitalcommons.lmu.edu/cgi/viewcontent.cgi?article=1007&context=econ_fac.
IPCC (Intergovernmental Panel on Climate Change). 2007. Climate Change 2007. Impacts, Adaptation and Vulnerability. Report from Working Group 2. https://www.ipcc.ch/site/assets/uploads/2018/03/ar4_wg2_full_report.pdf.
International Energy Agency. 2021a. "Global Carbon Dioxide Emissions are Set for Their Second-Biggest Increase in History—News." *IEA*, April

20. https://www.iea.org/news/global-carbon-dioxide-emissions-are-set-for-their-second-biggest-increase-in-history.
International Energy Agency. 2021b. "Global Energy Review 2021: Assessing the Effects of Economic Recoveries on Global Energy Demand and CO_2 Emissions in 2021." 36.
International Monetary Fund. 2022. "Navigating the High Inflation Environment." *Global Financial Stability Report*. October. https://www.imf.org/en/Publications/GFSR/Issues/2022/10/11/global-financial-stability-report-october-2022.
IPCC (Intergovernmental Panel on Climate Change). 2019. "Global Warming of 1.5 °C." https://www.ipcc.ch/site/assets/uploads/sites/2/2022/06/SR15_Full_Report_HR.pdf.
Ishiwatari, Mikio. 2021. "Institutional Coordination of Disaster Management: Engaging National and Local Governments in Japan." *Natural Hazards Review* 22 (1): 04020059. https://doi.org/10.1061/(ASCE)NH.1527-6996.0000423.
Janunta, Andrea. 2021. "Rising Global Temperatures 'Inexorably Closer' to Climate Tipping Point, WMO Says." *World Economic Forum*, May 27. https://www.weforum.org/agenda/2021/05/we-are-getting-closer-to-breaking-the-1-5-degrees-warming-threshold/.
Japan Meteorological Agency. 2013. "Emergency Warning System." https://www.jma.go.jp/jma/en/Emergency_Warning/ew_index.html.
Jorgenson Dale W., and Peter J. Wilcoxen. 1995. "The Economic Effects of a Carbon Tax." In *Shaping National Responses to Climate Change*, ed. Henry Lee, 237-60. Washington: The Island Press.
Kahn, Matthew E., Kamiar Mohaddes, Ryan N. C. Ng, M. Hashem Pesaran, Mehdi Raissi, and Jui-Chung Yang. 2019. "Long-Term Macroeconomic Effects of Climate Change: A Cross-Country Analysis." w26167. *National Bureau of Economic Research*. https://doi.org/10.3386/w26167.
Maya, C. 2016. "State Tops in Health-Care Spending." *The Hindu*, April 28, sec. Kerala. https://www.thehindu.com/news/national/kerala/state-tops-in-healthcare-spending/article8530723.ece.
Mohan, Matthew. 2021. "Transition to Cleaner Energy Will Involve 'Trade-Offs', EMA to Reduce Cost Impact Where Feasible: Chief Executive." *CNA*. https://www.channelnewsasia.com/singapore/singapore-climate-change-electricity-price-energy-transition-ema-2110516.
Morris, David Z. 2018. "Renewable Energy Surges to 18% of U.S. Power Mix." *Fortune*. February 19. https://fortune.com/2018/02/18/renewable-energy-us-power-mix/.
Motherway, Brian. 2020. "Energy Efficiency 2020." *Energy Efficiency*, 105.

Munich, R.E. 2022. "Natural Disaster Risks: Losses Are Trending Upwards | Munich Re." https://www.munichre.com/en/risks/natural-disasters-losses-are-trending-upwards.html.
Nahm, Jonas M., Scot M. Miller, and Johannes Urpelainen. 2022. "G20's US$14-Trillion Economic Stimulus Reneges on Emissions Pledges." *Nature* 603 (7899): 28–31. https://doi.org/10.1038/d41586-022-00540-6.
Official Monetary and Financial Institutions Forum. 2019. "Central Banks and Climate Change." *Global Public Investor.* https://www.omfif.org/wp-content/uploads/2020/02/ESG.pdf.
OECD. 2022. "Focus on Green Recovery." OECD. https://www.oecd.org/coronavirus/en/themes/green-recovery.
Park, Albert. 2022. Strengthening Connectivity for Sustainable and Inclusive Recovery. *Asian Development Bank*, May 24. https://unescap.org/sites/default/d8files/event-documents/Albert_Park_slides_roundtable_supply_chains.pdf.
Parry, M. 2009. "Closing the loop between mitigation, impacts and adaptation." *Climatic Change* 96: 23–27. August 15. https://doi.org/10.1007/s10584-009-9646-7
Rapier, Robert. 2019. "The World's Top 10 Carbon Dioxide Emitters." *Forbes*, December 4. https://www.forbes.com/sites/rrapier/2019/12/04/the-worlds-top-10-carbon-dioxide-emitters/.
Ritchie, Hannah. 2019. "How Do CO2 Emissions Compare When We Adjust for Trade?" *Our World in Data*, October 7. https://ourworldindata.org/consumption-based-co2.
Roy, Pinaki. 2014. "Climate Change: Bangladesh Lashed by Cyclones". *The Straits Times*, August 30. https://www.straitstimes.com/asia/south-asia/climate-change-bangladesh-lashed-by-cyclones.
Sachs, Jeffrey D. 2021. "Fixing Climate Finance." *Project Syndicate*, November 23. https://www.project-syndicate.org/commentary/fixing-climate-finance-requires-global-rules-by-jeffrey-d-sachs-2021-11
Segal, Stephanie. 2020. "Breaking down the G20 Covid-19 Fiscal Response." Center for Strategic and International Studies. https://www.csis.org/analysis/breaking-down-g20-covid-19-fiscal-response.
Stiglitz, Joseph E., and Nicholas Stern. 2017. "Report of High-Level Commission on Carbon Prices." *Carbon Pricing Leadership Coalition*, May 29.
The World Bank. 2016. "Bangladesh: Building Resilience to Climate Change." *World Bank*, October 9. https://doi.org/10/07/bangladesh-building-resilience-to-climate-change.
The World Bank. 2017a. "Philippines Launches Innovative Insurance Program to Boost Natural Disaster Risk Management." *World Bank*, August 15.

https://www.worldbank.org/en/news/press-release/2017/08/15/philip pines-launches-innovative-insurance-program-to-boost-natural-disaster-risk-management.

The World Bank. 2017b. "World Bank Resilience M&E (ReM&E): Good Practice Case Studies." *World Bank Group*. https://openknowledge.worldb ank.org/bitstream/handle/10986/28387/119939-WP-PUBLIC-P155632-28p-ReMECasestudiesFinal.pdf?sequence=1&isAllowed=y.

The World Bank. 2021. Vietnam: Climate Change Knowledge Portal. https:// climateknowledgeportal.worldbank.org/country/vietnam/vulnerability#:~: text=Vietnam%20has%20extremely%20high%20exposure,cyclones%20and%20t heir%20associated%20hazards.

The World Bank. 2022. "Key Highlights: Country Climate and Development Report for Vietnam." *World Bank*, July 1. https://www.worldbank.org/en/country/vietnam/brief/key-highlights-country-climate-and-development-rep ort-for-vietnam.

Thomas, Vinod. 2017. *Climate Change and Natural Disasters: Transforming Economics and Policies for a Sustainable Future*. 1st ed. New Brunswick (U.S.A.): Transaction Publishers [2016]: Routledge. https://doi.org/10. 4324/9781315081045.

Thomas, Vinod, and Namrata Chindarkar. 2019. *Economic Evaluation of Sustainable Development*. Singapore: Springer Singapore. https://doi.org/10.1007/ 978-981-13-6389-4.

Thomas, Vinod. 2022a. "Commentary: Why It Pays for Singapore to Be Much More Ambitious in Raising Carbon Tax." *CNA*. https://www.channelnewsa sia.com/commentary/carbon-tax-why-increase-singapore-budget-emissions-climate-change-2499051.

Thomas, Vinod. 2022b. "India's Climate Imperative." *The Hindu*, July 18. https://www.thehindu.com/opinion/op-ed/indias-climate-imperative/ article65654280.ece.

United Nations Economic and Social Commission for Asia and the Pacific-United Nations Office for Disaster Risk Reduction. 2012. Reducing Vulnerability and Exposure to Disasters. The Asia Pacific Disaster Report. https://www.unescap.org/publications/asia-pacific-disaster-report-2012-reducing-vulnerability-and-exposure-disasters.

United Nations Office for Disaster Risk Reduction. 2013. "Evacuation Saves Whole Island from Typhoon Haiyan." November 15. https://www.undrr. org/news/evacuation-saves-whole-island-typhoon-haiyan.

United Nations Office for Disaster Risk Reduction. 2021. https://news.un.org/ en/story/2021/09/1098662

United Nations Environment Programme. 2019. "Cut Global Emissions by 7.6 Percent Every Year for next Decade to Meet 1.5°C Paris Target - UN

Report." November 26. https://www.unep.org/news-and-stories/press-release/cut-global-emissions-76-percent-every-year-next-decade-meet-15degc.
US EPA (United States Environmental Protection Agency). 2001. "National Air Quality and Emissions Trends Report, 1999." 27. U.S. Environmental Protection Agency. https://www.epa.gov/sites/default/files/2017-11/documents/trends_report_1999.pdf.
Watts, Nick, Markus Amann, Nigel Arnell, Sonja Ayeb-Karlsson, Kristine Belesova, Maxwell Boykoff, Peter Byass, et al. 2019. "The 2019 Report of The Lancet Countdown on Health and Climate Change: Ensuring That the Health of a Child Born Today Is Not Defined by a Changing Climate." *Lancet* 394 (10211): 1836–78. https://doi.org/10.1016/S0140-6736(19)32596-6.
Wei, Y.M., R. Han, C. Wang, et al. 2020. "Self-preservation Strategy for Approaching Global Warming Targets in the post-Paris Agreement Era." *Nature Communications* 11: 1624. https://doi.org/10.1038/s41467-020-15453-z.
World Meteorological Organization. 2020. "World's Deadliest Tropical Cyclone Was 50 Years Ago." November 12. https://public.wmo.int/en/media/news/world%E2%80%99s-deadliest-tropical-cyclone-was-50-years-ago.
World Meteorological Organization. 2021. "Greenhouse Gas Bulletin: Another Year another Record." October 21. https://public.wmo.int/en/media/press-release/greenhouse-gas-bulletin-another-year-another-record.
Yeo, Sophie. 2019. "Climate Finance: The Money Trail." *Springer Nature Limited*, September 19.
Yusuf, Shahid. 2011. *Cities as Engines of Growth | Frontiers in Development Policy*. https://elibrary.worldbank.org/doi/10.1596/9780821387856_CH23.
Yuzal, Hendri, Karl Kim, Pradip Pant, and Eric Yamashita. 2017. "Tsunami Evacuation Buildings and Evacuation Planning in Banda Aceh, Indonesia." *Journal of Emergency Management (Weston, Mass.)* 15 (1): 49–61. https://doi.org/10.5055/jem.2017.0312.

CHAPTER 9

Transformative Change

The size of your dreams should always exceed your current capacity to achieve them. Ellen Johnson Sirleaf

Recognising just how extreme the danger of climate change is and how it is reshaping life are paramount for mapping a way forward out of the crisis (Mundy 2022; Smil 2022). Climate change dramatises the reality that resilience needs to meet higher risks by re-building smarter and better. The imperative to recognise inflection points and anticipate bigger risks also often goes with the needed readiness to opt for transformational solutions (Grove 1999). This scenario may also be present in pandemics, cyber threats, and geopolitical tension. But the climate crisis stands out as a game stopper, since it is already late to prevent massive damage and climate measures must now focus on avoiding catastrophes.

This chapter examines the implications of this conclusion. It is no longer enough to make gradual changes in favour of sustainable development. Sticking with incremental changes favoured by economists is already proving to be a recipe for disaster. Economies need to make transformative changes in how they value the environment to make progress in achieving a sustainable pattern of economic growth and to promote wellbeing. Because climate change's deep historical antecedents and roots in a global political economy are hard to unwind, far-reaching reforms are needed to find a better direction.

To be clear, humanity will land in a new equilibrium, as it always does. The question is if it can be a reasonably soft landing or if it would have to be an unacceptably hard landing. Settling on new trajectories forced upon economies by past policy inaction and mounting calamities would not be an outcome to hope for. Several high income countries, despite their legacy of wasteful consumption and carbon-intensive investment, have reduced emission-GDP ratios through structural changes, greater efficiency, and a switch to cleaner fuels. But projections of damages have also proved to be too conservative. In light of this, climate policy may need to overshoot the current targets to avert worst-case scenarios. At the same time, misfortunes also open opportunities to act which can spur more desirable outcomes. The jump in oil prices triggered by Russia's war in Ukraine could be a boost to the attraction of investing massively in renewable energy and walk away not only from coal but also from oil and gas. Increases in carbon prices in carbon markets can underpin such a switch.

The Big Picture

With the rise in environmental calamities, it is vital to be aware of tipping points and irrecoverable damage, as Chapter 5 alerted. If a threshold is crossed, there is no coming back—consider the cases of the threatened extinction of the Javan rhino, Sumatran orangutans, and the hawksbill turtle. Climate tipping points are coordinates of no return in the sense that changes already occurring in the climatic system keep perpetuating themselves. These shifts carry with them a sense of irreversibility and extreme danger. Rapidly melting ice sheets in the North Pole are affecting lives and livelihoods, and the financial viability of countries. For example, precariously rising sea levels render countries in the Caribbean Sea, including locations like Panama's San Blas islands, particularly vulnerable.

In this context, countries with Arctic territories, including Sweden, Finland, and Norway, have warned for some time that irreversibility of rapidly melting ice sheets because of global warming may soon become a reality because of little to no global climate action (Doyle 2007). Indeed, the UN's climate science panel in 2019 warned that Arctic warming doubled that of the global mean, which could bring the earth closer to the tipping point of rising sea levels (France 24 2022). McKay et al. (2022), among other studies, highlights that the planet is on a 2 to 3 °C of global warming path. Even if Paris Agreement pledges are met, warming will approach 2 °C. IPCC notes with "high confidence" that "soft limits to some human adaptations have been reached ... and hard limits to adaptation have been reached in some ecosystems ... and additional human and natural systems will reach adaptation limits" (IPCC 2022). The bottom line is that the need to adapt to extreme conditions is now a foregone conclusion.

Adaptation holds greater political appeal for countries than mitigation because a country's citizens can directly capture its benefits. Some countries may not appreciate that their investment in mitigation could benefit others that may not be making similar investment. Even so, investments in adaptation are still grossly inadequate for dealing with the emerging eventualities, partly because investors often find country efforts in this respect not all that bankable, something that can change enormously with policies. To better see the implications of inadequate preparation, and of the uncertainty of risk scenarios, it pays to build in system stress tests in health care and physical infrastructure, among other areas, and to do these tests regularly—just as central banks do for financial systems. Tackling the extent of vulnerability of people on low incomes, which these tests may reveal, is essential as it identifies weakest links in the population chain.

Examining the confluence of risks, including pandemics and geopolitical threats, points to the need to go beyond adapting to the realities to also building far greater resilience to emerging dangers. UNDRR has a sobering conclusion: "Despite progress, risk creation is outstripping risk reduction" (UNDRR 2022). Clearly, the juxtaposition of growing dangers from global warming with the continuing threats associated with the COVID-19 pandemic and geopolitical conflicts heightens the daunting challenge of sustainable development. Resilience mechanisms and systems need to be "rewired". The need is clear for transformative changes in how vulnerability, exposure, and hazards are viewed.

Climate resilience is not only about greater coping in the form of adaptation but also mixing with greater preparedness in the form of mitigation. As underscored in earlier chapters, when the threat combines exogenous and endogenous dimensions, as in global warming, the emphasis needs to be not only adaptation to the inevitable but also on mitigation to ameliorate the risk. The same applies to pandemics and financial crises. But society has especially underinvested in mitigating risks like climate change. It is time to change that course.

Global warming scenarios also show the big gains that can be had from prompt responses (IPCC 2021). Delaying policies and investments will come at a high price—by increasing the cost of future measures and making it more necessary to take even more extreme steps. Climate policy needs to overcome past neglect, and large carbon-emitting nations must resist the special interests clinging to carbon-intensive energy. This is a chance to restart development in a cleaner environment left by the COVID-19 pandemic and for "big bang" climate restructuring by governments and businesses that could be more cost-efficient than gradual changes (Way et al. 2022). It helps that investing heavily in emergency situations gained acceptance during COVID-19. Despite the worries of a global economic downturn, COP27 and beyond have a chance to move the needle on both mitigation and adaptation as well as climate finance. The pickup in investments in renewables, generated by shortages and high prices of oil and gas, could give a boost to mitigation plans.

Unfortunately, economic interests are stacked against recognising this, let alone formulating policies for protecting nature. Rewards are loaded in favour of short-term profits that can accrue from destructive and value extracting operations rather than regenerative and value-creating GDP growth (Mazzucato 2018). Mainstream economics is shielding these patterns by not recognising the value of investments in conserving nature, or ones whose benefits accrue over the horizon (Stern et al. 2021). The escalating damage from global warming, population pressure, and biodiversity loss call into question this approach. In the face of this emerging threat, a paradigm shift is needed to prioritise building resilience that includes making space for innovation and risk-taking in the search for solutions.

It is indeed a problem that the priorities set in the economic analysis of short-term growth underplay the requirements of sustainable development. This neglect feeds the mismatch between the time horizons in policymaking and the obligations of sustainable climate policy (Boston and Lempp 2011). Policymakers typically survive and prosper in a world where shorter horizons dominate based on perceived gains, including the will of powerful interest groups. Public opinion in favour of sustainable development could influence this mismatch of time horizons. That is why this book has stressed the need to scrutinise the linkages and identify the sources of the crisis.

Triage Financing and Pricing

The mismatch between needed measures and the priorities set is best illustrated by the gaps in climate financing. The struggle to reach the UN's US$100 billion annual target for climate financing for developing countries shows how far off governments and businesses are in prioritising resources to fight climate change. Resources also need to be managed effectively because the bar for risks is being dramatically raised. Country experiences show that the efficiency in the use of resources raised for environmental protection makes it easier to generate more funds for this purpose.

Historically, developed countries have, by far, led the charge in carbon-intensive growth. A few countries and localities have taken the initiative to use COVID-19 fiscal and economic stimulus packages as an opportunity to advance environmental sustainability and green growth. The UK announced in May 2022 a plan to spend GBP200 million for new walking and cycling schemes across England (gov.uk 2022). In July 2021, the European Commission adopted proposals to make the EU's policies on the environment, energy, transport, and taxation consistent with lowering GHGs by at least 55% by 2030, compared with 1990 levels (Cifuentes-Faura 2022). As part of a green recovery, Tasmania targeted 200% clean energy, including energy exports, by 2040 (Morton 2020). The C40 Cities made up of 97 cities that together comprise one quarter of the global economy pledged environmental sustainability in their COVID-19 economic recovery plans (C40 2022, 40). Australia passed a major legislation mandating the reduction of carbon emissions by 43% from 2005 levels by 2030 and reach net zero emissions by 2050 (Langley 2022). In August 2022, the US Senate, which has historically lagged on substantive climate action, finally passed legislation, the Inflation Reduction Act, dedicating US$370 billion to climate and energy investments to make a radical cut in greenhouse gas emissions by 2030 (Box 9.1). These are positive developments, but the challenge will be to scale up such commitments and, crucially, to ensure that the implementation of interventions being funded by these investments is timely and efficient.

> **Box 9.1 The 2022 US Climate, Health, and Tax Bill**
>
> A bill passed by the Senate and signed into law by the President sets aside US$370 billion for climate and energy investments. The bill aims to help the US achieve its 40% reduction in GHG emissions goal by 2030. Measures include tax incentives, resourcing for alternative energy sources, and funding for smaller or disadvantaged communities to build resilience against climate change.
>
> Vice President Kamala Harris cast the tie-breaking vote in favour of the bill. If well-implemented, it would be the US' most impactful climate law in history. Some might criticise this move as being too little, too late, with the Democrats and Republicans traditionally at loggerheads, thus slowing down decisive climate action and limiting the political will needed to push through with such changes. A highly conservative Supreme Court has delivered a blow to the ability of the Environmental Protection Agency to advance environmental reforms. If a less pro-climate President were to be elected in 2024, it will hurt the implementation of this landmark bill.
>
> Yet, the US$370 billion is meaningful, especially in comparison with the US 2021 defence budget of US$754 billion (Peterson Foundation 2022). At almost half of the defence spending bill, this shows the renewed commitment of the US towards substantive climate action. That said, if climate investment can be treated with the same priority and intensity as defence investment, it is imaginable that the positive impacts on climate change can be much more. Most pertinently, if all countries were to commit a similar proportion of their budgets to climate investment, results would be much more achievable.
>
> *Source* Cochrane (2022).

A part of getting the most from financing relates to the motivation that drives low-carbon growth. While the experience with carbon pricing, discussed in Chapter 8, is growing, far more needs to be done in price policy to come close to the pricing levels that would achieve a significant lowering of emissions. An IMF proposal for the world's largest emitters to set a floor price of US$25–US$75 per tonne of carbon, differentiated by country income levels, would be a place to begin (Parry et al. 2021). All countries should start using carbon pricing, even if its rate is differentiated, say in two or three tiers, according to the per capita income levels of countries.

In addition to market forces, regulation of the command-and-control variety, as seen in phases of COVID-19, is needed to tackle climate change. The guiding hand of government usually suffices for well-functioning markets. But with out-of-control climate change, the extreme failure of markets reflected in societal damages from economic activity could prompt the heavy hand of governments and businesses to reign in emissions—especially as time is running out.

ENERGY TRANSITION

Energy, responsible for about three-fourths of GHGs in the air, is the focal point of the low-carbon transformation that is needed worldwide. The growing demand for energy is a huge obstacle to achieving this transition challenge.

Over 733 million people do not yet have access to electricity and nearly 2.4 billion are not cooking with clean fuels (International Energy Agency 2022a, b). Electric power has made progress in adopting renewables in its energy mix, but a far bigger switch from fossil fuels is needed for domestic heating and cooling to lower effluents (REN21 Secretariat 2022). Industry, the largest energy user, and transport are making disappointing progress in switching to renewables. A UNEP report concludes that the "slow progress in energy conservation, energy efficiency and renewables prevent the transition away from fossil fuels …" (UNEP 2020).

International Energy Agency gives a thoughtful framework for net zero emissions by 2050 centred on policies to meet the rising energy demand (IEA 2021). For a transition to net zero, substantial measures are needed on multiple fronts. The single biggest breakthrough needed will be a much faster deployment of clean energy technologies. The infrastructure for renewable energy in terms of development, storage, and distribution, including advanced batteries, needs to be vastly improved. Consumers, for their part, need to be convinced of the value of switching to renewable sources. Energy security needs to be assured, especially for the vulnerable.

Calls were made in 2022 for taxing fossil fuel companies both on the principle that polluters must pay and on grounds of equity. Furthermore, IEA's analysis shows that the huge financial windfall for oil and gas resulting from high energy prices linked to COVID-19 and Russia's invasion of Ukraine in early 2022 could be directed to clean energy investment—"a once-in-a-generation opportunity" (Birol 2022). UN Secretary General António Guterres has urged governments to levy windfall taxes on oil and gas companies and to use the proceeds to provide some protection for those who are vulnerable to climate calamities and food and energy price hikes. The European Commission has proposed a temporary tax on the "surplus profits" on fossil fuel producers to help compensate for rising utility costs. A 2022 IMF paper proposes a permanent tax on windfall profits from the extraction of fossil fuels (Baunsgaard and Vernon 2022).

The spike in 2022 energy prices from the war and COVID-19, is causing considerable hardships for consumers globally. Of greatest concern are the availability and prices of basic energy for the vulnerable population, the elderly, and the poor. Worrying would be an increase in demand for energy from coal that would set back the decarbonisation agenda. In the short term, it is understandable to make a case for using existing fossil fuel stocks without adding to capacity and better use of existing capacity of cleaner categories of non-renewables. In any event, the pressure on energy prices from geopolitical conflicts and the pandemic ought not to be conflated with the pursuit of climate sustainability.

At the same time, the rise in prices of oil and gas—as well as carbon prices in 2022–23—presents incentives to meet energy demand through alternatives like renewables. Indeed, renewable energy technologies and electric vehicles are seeing an upsurge in the wake of oil price increases (IEA 2022a, b). Russia's invasion and the rise in natural gas prices have held up coal demand, but the expansion of renewables because of changes in relative prices has been dominant. It is essential that renewables are adopted on a large enough scale to avert further energy shortages. A massive boost in renewable plant capacity and energy generated by renewable energy as well as trading renewables across national boundaries are called for.

The energy–climate interaction presents tough policy choices. Because of the existential nature of the climate crisis, achieving zero carbon growth will need to be the top policy priority for all nations, even if it means less economic growth as measured in GDP terms for the short term. There are financial costs of achieving zero carbon growth, and lower-income countries which are far from realising this goal will only make headway if sizeable carbon finance is made available to them. Furthermore, climate investment requires resources. Economic growth is a source of generating resources, but if it is based on high carbon, the process will aggravate the climate crisis and therefore be self-defeating. The way forward is for economies to shift the focus away from GDP-centric growth and towards low carbon, quality growth.

Energy security is a huge priority for sustaining a drive to transition out of fossil fuels in all countries regardless of their income categorisation. A 2022 IMF paper presents a scenario where increases in the share of nuclear, renewable, and other non-hydrocarbon energy, coupled with higher energy efficiency, could bring about both a drastic cut in carbon effluents and greater energy security in Europe (Cevik 2022). And it seems possible that continual innovation and investment will be able to drive renewable energy to the desired scale (Krugman 2021). But innovation and investment in renewable energy will require forceful policies and reform shifts out of hydrocarbons, as well as steps to explicitly lower energy dependence and susceptibility to energy price volatility.

The IEA pathway to net zero emissions in 2050 sees no new oil and gas fields and coal mines being developed beyond those already approved by 2021. Governments need to take decisive steps to deal with the powerful fossil fuel interest groups that unethically campaign against climate and anti-pollution policies. Globally, the coal, oil and gas industries still receive various forms of support from governments (Carrington 2019; Irfan 2019). These subsidies should be going into increasing the capabilities and capacities of renewable energy firms. Time is of the essence. Clean energy should halve carbon emissions by 2030 and enable net zero by 2050. It would be a game changer if innovative breakthroughs such as zero carbon power through fusion reaction can advance with the lightning speed that characterised the finding of a vaccine for COVID-19. Their costs need to come down drastically. Societal and individual behaviour and pressure for action must change if the government and private sector intervention in this respect are to be sustained. Nature-based solutions have vast advantages of relying on regenerative capital but struggle to go to scale and be commercially viable. Green solutions need the support of the public to be widely adopted.

GLOBAL ECONOMIC POLICYMAKING

At the end of the day, debates and discussions on sustainability boil down to socioeconomic and environmental legislation, and policy measures countries and the global community take. Without seeking to be exhaustive, the following seven bottom lines can be considered within the frameworks offered in this book. They set out the problem and a specific response within the realm of socio-economic policy, illustrative of what can be achieved:

- Systematic dislocations like climate change call for a change in the traditional understanding of risk as just a departure from the norm; they also qualify resilience from getting back to normal to anticipating rising risks and re-building better.
  - *In such a scenario, countries would want to do regular stress tests of their resilience capabilities in the face of rising risks.*
- Scientific evidence obliges us to shift from thinking of climate change as a pure act of nature to seeing it also as human-induced.
  - *Accordingly, prevention should become as big a priority in country investment as simply coping.*
- Spillovers should become central to growth economics and economists should take greater note of the scientific evidence on risk and resilience.
  - *Encouraging GDP growth at any cost, like subsidising high carbon growth, must stop everywhere.*
- Public backing for tackling climate change depends on how well scientists, economists, and the media link global warming, weather disasters, and people's well-being.
  - *Meteorologists must go beyond simply describing weather disasters to linking these disasters to rising carbon emissions.*
- Immediate inconveniences seem to take precedence over emerging catastrophes, and projections for 2100 wrongly give the impression of climate change as primarily a distant threat.
  - *While the need for energy security is uppermost among priorities, meeting energy demand must be made contingent on decarbonisation without delay.*
- Costs will be incurred in dealing with the harm from burning fossil fuels which places a premium on innovation and risk-taking enabling the low-carbon transition, especially by the private sector.

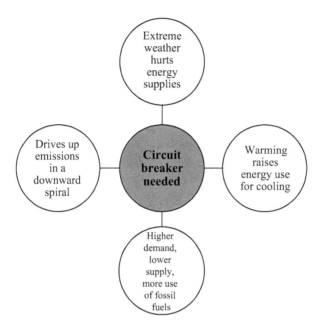

Fig. 9.1 Aggravation of the climate crisis (*Source* Author's depiction)

– *Rich countries must provide far more carbon finance to developing countries, and, at the same time, invest heavily in innovations in renewable energy, carbon capture and sequestration.*

- Climate change can present a non-convergence of the problem, as severe weather can destabilise energy supplies, which coupled with demand for more cooling, prompts more fossil fuel use, driving up effluents (see Fig. 9.1).

– *This could call for circuit breakers that stop the downward spiral, and coordination among countries to help avert a catastrophe.*

Societal and Individual Behaviour

Policy measures are rooted, in good measure, in people's values, behaviours, and mindsets. The massive scale of government intervention to fight the COVID-19 pandemic and the groundswell of public support for this show that, faced with extreme danger, countries can act rapidly and collectively. That resolve, albeit in far greater measure, is now required to decisively tackle the climate crisis.

For this to be successful, societal and individual behaviour must change in sync with government intervention. The critical point to get people fully behind the need to deal with climate crisis has already been reached—as manifested by the collective shudder to successive seasons of devastating forest fires and the sharp increase in extreme floods and storms. Public awareness, already traumatised by the COVID-19 pandemic, could provide the needed push for governments, especially in the large-polluting countries, to do what it takes to avert a series of catastrophes that will affect generations.

But acute awareness of the threats does not imply the avoidance of every contingency and risk (Quah and Thomas 2022). Flexibility in how resilience is built matters. The COVID-19 lockdowns, for example, proved to be an economically costly solution to tackling the pandemic. By the same token, closing polluting powerplants to cut carbon emissions is more costly than using low-carbon fuels or putting high taxes on emissions. Using resources sustainably calls for ways and means to manage inherent conflicts for the common good (Walljasper 2011).

Greater resilience is not about simply reinforcing robustness to withstand bigger shocks (Brunnermeier 2021). Large rigid skyscrapers do not last long; those that do are the ones that can sway. In the same way, large shocks do not need to be fought with great strength, but rather with flexible and adaptive interventions. COVID-19 showed that financial strength alone was no defence, especially because the virus was constantly mutating. This taught all countries that flexibility in how resilience is built matters. But clear government leadership and civic cooperation are needed, as comparative studies of East Asian countries suggest (Ma et al. 2021).

A takeaway from COVID-19 is how building resilience of health systems shapes risk profiles related to the pandemic. The extent of damage to socioeconomic systems depends on the readiness already built in, which can limit shocks and speed recoveries. The recovery path from COVID-19, both its speed and breadth, is determined by the resilience added during the pandemic—and here lies the value of building capabilities. Declines in health service use during the pandemic could have important effects on population health. More generally, the time of the pandemic can also see declines in health service affecting the health of the population, some of which may have preceded COVID-19 (Arsenault et al. 2022).

The Quality of Growth

The overriding challenge is to get away from targeting the quantity and pace of GDP and to switch to a focus on the *quality* of growth that includes environmental sustainability and social inclusion in addition to economic output (Munasinghe 2019; Raworth 2017; Thomas et al. 2000; World Bank 1992). As previous chapters mentioned, GDP is not a good measure of progress encompassing socioeconomic and environmental dimensions (Stiglitz and Durand 2019). Development is about improving the quality of life in ways that are sustainable environmentally, socially, and economically. The economic aspirations of society need to be fulfilled in ways in which humankind can live in harmony with nature, as envisioned in the UN's SDGs (Sachs 2015). This means investing not only in physical and social capital but also in natural capital.

Indeed, even in formulations for improving incomes sustainably, all three forms of capital—physical and financial, social and human, and environmental and natural capital—are involved. While attention to all these forms is needed for sustainable development, interest has focused on physical capital, as reflected in investment in this form of capital. Increased attention is being paid to human capital, a trend that has become apparent in recent years. Natural capital, however, still gets scant consideration, so much so that the trend has been to disinvest rather than invest in natural capital.

There is great merit in thinking of a circular economy that minimises waste, and stresses regenerative rather than extractive growth. Production and distribution of goods and services would be accompanied by collection, reuse, and recycling that minimises residual waste. A circular economy would motivate reusing products, rather than wasting them and drawing on new resources. Waste would be returned to the economy providing multiple benefits, including protecting biodiversity (Kweku et al. 2022).

As noted, GDP measures, by not including societal damages, give a misleading picture of economic progress. Several efforts try to adjust for this lapse, especially by accounting for natural capital depletion. The World Bank's *Changing of Wealth of Nations* shows the sum of produced capital, human capital, and natural capital for 146 nations (The World Bank 2021). It finds that during 1995–2018 this measure of the world's wealth grew about 190% or 2.9% a year, in comparison with an estimated GDP growth of 3.2% a year. This calculation does not account for global externalities and damages, especially climate change. In a different approach and time period, UNEP's *Inclusive Wealth 2018*, the wealth for 135 countries was 44% higher in 2014 over 1990 implying an increase of 1.8% a year in contrast to GDP growth estimated at 3.4% annually (UNEP 2018). The US White House Office of Science and Technology is initiating work on developing a natural accounts framework for the country (The White House 2022).

The argument is often made that the single-minded focus on economic growth has been good for poverty reduction. But it should be noted that most formulations of this position represent a tautology. Economic growth is measured by GDP growth, and poverty is defined by GDP-based income poverty. Therefore, unless income distribution changes (which takes time), GDP growth and income poverty reduction naturally go together. All the problems of GDP measurement remain. Higher incomes are associated with the physical ingredients of well-being, but then the quality of growth decides if it will be sustainable and welfare-enhancing or not.

Academic curricula in economics and business schools need to be overhauled to reflect this change in development objectives that can better deliver sustainable development. A huge premium is to be had in achieving more regenerative growth, for example, using renewable energy, rather than destructive growth, for example, resorting to deforestation. Decarbonisation, however, does not mean curtailing economic activity, but it does call for reducing carbon-intensive energy by promoting renewable energy, which is now a competitive alternative, and expanding pollution-control technologies (Thomas 1980). Both can be done at a fraction of the cost of limiting economic activity (Ambrose 2019; Hamer 1976).

New technologies can play a huge role in making growth more sustainable, and there has been striking technological progress, for example, green hydrogen that can be a clean form of energy, and carbon capture from the air (Kane and Gill 2022). Both have the potential of slowing and even reversing climate change. But investment in new technologies needs to speed up, taking into account both the fact that the stock of pollution already in the air determines outcomes in the coming decades, and that there are serious lead times for these approaches to bear fruit commercially. The demand for hydrogen has risen sharply in recent years from very low levels and is projected to grow 5–6 times by 2050. But some reports say that most of the current hydrogen production is based on fossil fuels, and not much is derived from clean sources. By one projection, carbon capture can bring about 14% of the GHG emissions reduction aimed for in 2050 (C2ES 2020). Rising prices of carbon in recent years could make carbon capture increasingly profitable (Dans 2022). It would not be a panacea, but a valuable complement to other decarbonisation measures.

Policy should be motivated by knowing that climate change involves investment that will, if done right, generate sustainable economic growth and employment. COVID-19 stimulus spending, it should be noted, was a short-term boost to consumption. While the economic slump caused by COVID-19 resulted in airborne toxic chemicals to decline, carbon dioxide, the chief culprit in global warming, returned to previous levels as early as in the pandemic's first year (2020), and emissions were rising in 2022. Global warming also accelerated in that period from past accumulation (UNEP 2020; Lindsey 2020).

The merits of focusing on the quality rather than the quantity of growth is shown by the social cost–benefit analyses of climate investment. The benefits of implementing the Paris Agreement, for example, far outweigh the costs of inaction (WHO 2018). The intuition and numbers on the side of making transformative change are clear. The consequences of not adopting far-reaching changes are dire. But the big question is the political will of big-polluting nations to take on the special interests still stuck on polluting forms of energy (Ziady 2020).

For the signals being provided for policy, it matters what the constituents are for measures of progress. Various indexes track environmental status, but most measures of economic competitiveness and economic betterment neglect environmental and climate damage (Picciotto and Thomas 2021). An important exception is the UN's Planetary Pressures-Adjusted Human Development Index (PHDI) that includes planetary impacts per person to an existing human development index (UNDP 2022; Thomas 2021). This adjustment typically lowers the rankings of many high-income countries, because of their high per capita consumption and carbon intensity. For example, Norway fell 15 places in the PHDI after placing first place out of 169 countries in the 2019 index because of adjustments made for planetary pressures per person. The US fell 45 places from the 16th.

Only a few industrialised countries have successfully decoupled economic growth from carbon increases in the past two decades (Cohen et al. 2018). Germany and the UK have each reduced emissions by 20% while growing their economies. However, the developing economies of China and India grew in the same period amid rising carbon emissions. Russia, Brazil, and Saudi Arabia, too, saw rising emissions. Many countries remain indifferent and some are even hostile to bold climate plans. This must change.

Informed by Evaluation

Because risks are dynamic, projects and programmes for resilience building should be capable of being evaluated and adjusted midstream. The evaluation of climate policy needs to be rigorous, flexible, and timely. Cost–benefit analysis can be used to evaluate sustainability with frameworks that attempt a quantification of non-market categories and that incorporate market interventions, such as carbon taxes, to account for externalities, such as pollution and congestion (Thomas and Chindarkar 2019).

Impact evaluation tries to assess causality and find attribution for the results obtained. In certain cases, they can be useful for assessing the effects of risk- reduction strategies. The application of new forms of impact evaluation is found to be useful, such as process tracing, systems mapping, and qualitative comparative analysis. All these use methods and tools that are rigorous, but they are not all necessarily experimental in design. Randomised controlled trials have had difficulty in tackling sustainable development over generations.

To better value natural capital, green accounting methods are increasingly available (Hamilton 2016). When the destruction of natural capital is not accounted for, the results inflate growth prospects (Dasgupta 2010, 2021). And not accounting for this sends the wrong signals for pursuing GDP growth at the expense of running down natural capital, which eventually hurts the growth process itself. Promising would be the use of the System of Environmental-Economic Accounting (SEEA) which tries to integrate economic and environmental information to give a fuller picture of economic-environmental interactions (UN 2014). Getting adequate data is usually a constraint in applying these valuations, but this can be overcome with innovative approaches.

Social impact analyses bring out the power of participatory processes in planning and implementation. Social and environmental impact assessments include the processes of analysing, monitoring, and managing intended and unintended social and environmental consequences, both positive and negative. Taking these evaluations seriously can make a big difference to decision-making as well as mid-course adjustments to projects already decided on and in execution. Mechanisms for enforcing environmental and social safeguards are essential in ensuring that development projects do not have harmful social and environmental effects. In the presence of negative externalities, the needed regulations to avoid damage will not be followed by the private sector or even public agents without environmental and social safeguards. A rigorous framework of evaluation can mix quantitative and qualitative methods depending on the context and the issues.

Conclusions

Growing dangers from global warming along with the continuing threats associated with the COVID-19 pandemic, geopolitical conflicts, global inflation, and food shortages highlight the need to shift from a preoccupation with short-term GDP growth that ignores socio-environmental damages to a focus on sustainable development. In the context of risk and resilience, vulnerability, exposure, and intensity (of hazards) need to be viewed as heavily anthropogenic and subject to responses made. Resilience mechanisms and systems need to be understood and practised as dynamic instruments for change. Scientific breakthroughs for clean and sustainable growth will help, as they have in many respects.

Policy responses will be aided by confronting the economic interests that are stacked against transformational change because they could hurt these interests even while producing net societal gains. Market instruments for pricing policies and the provision of financing are vastly underutilised and should be scaled up. Such a direction needs to be underpinned by changes in attitudes in favour of sustainable development and a preference for the quality of growth over short-term GDP expansion. COVID-19 and the 2008 global financial crisis demonstrate the capability of society to deal with wicked problems.

COP27 undoubtedly faces tough challenges in forging climate commitments, especially amid other global crises. But the greater realisation of the climate threat from a high carbon path, and the new possibilities for adopting renewables, should present important opportunities too. While climate answers have been elusive, some adversities can be a chance to drive new directions, for example, high oil prices motivating a far-reaching shift to cleaner fuels. The realisation of great risks can be an opportunity for changing behaviour in favour of resilience building based on scientific evidence. Such steps importantly include making space for innovations as well as taking risks in the search of solutions that might not have been anticipated.

Bibliography

Ambrose, Jillian. 2019. "Renewable Energy to Expand by 50% in Next Five Years—Report." *The Guardian*, October 21, sec. Environment. https://www.theguardian.com/environment/2019/oct/21/renewable-energy-to-expand-by-50-in-next-five-years-report.

Armstrong McKay, David I., Arie Staal, Jesse F. Abrams, Ricarda Winkelmann, Boris Sakschewski, Sina Loriani, Ingo Fetzer, Sarah E. Cornell, Johan Rockström, and Timothy M. Lenton. 2022. "Exceeding 1.5 °C Global Warming Could Trigger Multiple Climate Tipping Points." *Science (New York, N.Y.)* 377 (6611): eabn7950. https://doi.org/10.1126/science.abn7950.

Arsenault, C., A. Gage, M.K. Kim, et al. 2022. "COVID-19 and Resilience of Healthcare Systems in Ten Countries." *Nature Medicine* 28: 1314–24. https://doi.org/10.1038/s41591-022-01750-1.

Baunsgaard, Thomas, and Nate Vernon. 2022. *"Taxing Windfall Profits in the Energy Sector" IMF Note 2022/00X*. Washington, DC: International Monetary Fund.

Birol, Fatih. 2022. "What Does the Current Global Energy Crisis Mean for Energy Investment?" *IEA*, May 13. https://www.iea.org/commentaries/what-does-the-current-global-energy-crisis-mean-for-energy-investment.

Boston, Jonathan, and Frieder Lempp. 2011. "Climate Change: Explaining and Solving the Mismatch Between Scientific Urgency and Political Inertia. *Accounting, Auditing & Accountability Journal*, October 25. https://www.researchgate.net/publication/227429000_Climate_change_Explaining_and_solving_the_mismatch_between_scientific_urgency_and_political_inertia.

Brunnermeier, Markus K. 2021. *The Resilient Society*. Princeton Economics.

C40. 2022. "About C40." C40 Cities. 2022. https://www.c40.org/about-c40/.

Carrington, Damian. 2019. "Just 10% of Fossil Fuel Subsidy Cash 'Could Pay for Green Transition.'" *The Guardian*, August 1, sec. Environment. https://www.theguardian.com/environment/2019/aug/01/fossil-fuel-subsidy-cash-pay-green-energy-transition.

Cevik, Serhan. 2022. "Climate Change and Energy Security: The Dilemma or Opportunity of the Century?" *IMF Working Papers*, no. 2022/174 (September): 20.

Cifuentes-Faura, Javier. 2022. "European Union Policies and Their Role in Combating Climate Change over the Years." *Air Quality, Atmosphere & Health* 15 (January): 1333–40. Springer Nature B.V.https://doi.org/10.1007/s11869-022-01156-5.

Cochrane, Emily. 2022. "Senate Passes Climate and Tax Bill after Marathon Debate." *The New York Times*, August 7, sec. U.S. https://www.nytimes.com/live/2022/08/07/us/climate-tax-deal-vote.

Cohen, Gail, João Tovar Jalles, Prakash Loungani, and Ricardo Marto. 2018. "The Long-Run Decoupling of Emissions and Output: Evidence from the Largest Emitters." *IMF*. https://www.imf.org/en/Publications/WP/Issues/2018/03/13/The-Long-Run-Decoupling-of-Emissions-and-Output-Evidence-from-the-Largest-Emitters-45688.

C2ES. 2020. Carbon Capture. Center for Climate and Energy Solutions. https://www.c2es.org/content/carbon-capture/.

Dans, Enrique. 2022. Why Carbon Capture Could be a Profitable Business. *Forbes*, June 26. https://www.forbes.com/sites/enriquedans/2021/06/26/why-carbon-dioxide-capture-could-be-a-profitable-business/?sh=3c13240565fe.

Dasgupta, Partha. 2010. "The Place of Nature in Economic Development." In *Handbook of Development Economics*, vol. 5, 4977–5046. Elsevier. https://doi.org/10.1016/B978-0-444-52944-2.00012-4.

Dasgupta, Partha. 2021. *The Economics of Biodiversity: The Dasgupta Review: Full Report*. Updated: 18 February 2021. London: HM Treasury.

Doyle, Julie. 2007. Picturing the Clima(c)Tic: Greenpeace and the Representational Politics of Climate Change Communication. *Science as Culture* 16 (2): 129–50. https://doi.org/10.1080/09505430701368938.

France 24. 2022. "Arctic Warming Four Times Faster than Rest of Earth, More than Projected." *France 24*, August 11. https://www.france24.com/en/europe/20220811-arctic-warming-four-times-faster-than-rest-of-earth-much-higher-than-projections.

gov.uk. 2022. "Healthy, Cost-Effective Travel for Millions as Walking and Cycling Projects Get the Green Light." *GOV.UK*, May 14. https://www.gov.uk/government/news/healthy-cost-effective-travel-for-millions-as-walking-and-cycling-projects-get-the-green-light.

Grove, Andrew S. 1999. *Only the Paranoid Survive: How to Exploit the Crisis Points That Challenge Every Company*. Currency.

Hamer, John. 1976. "Pollution Control: Costs and Benefits." In *Editorial Research Reports 1976*, 145–64. *CQ Researcher Online*. Washington, D.C., United States: CQ Press. http://library.cqpress.com/cqresearcher/cqresrre1 976022700.

Hamilton, Kirk. 2016. Measuring Sustainability in the UN System of Environmental-Economic Accounting. *Environmental and Resource Economics* 64 (1): 25–36. https://doi.org/10.1007/s10640-015-9924-y.

International Energy Agency. 2022a. World Energy Outlook 2022a. October 19. https://www.iea.org/news/defying-expectations-co2-emissions-from-glo bal-fossil-fuel-combustion-are-set-to-grow-in-2022-by-only-a-fraction-of-last-year-s-big-increase.

International Energy Agency. 2022b. Tracking SDG 7. The Energy Progress Report 2022. https://trackingsdg7.esmap.org/data/files/download-docume nts/sdg7-report2022-executive_summary.pdf

International Energy Agency. 2021. "Net Zero by 2050: A Roadmap for the Global Energy Sector." France: IEA Publications. https://iea.blob.core. windows.net/assets/deebef5d-0c34-4539-9d0c-10b13d840027/NetZeroby 2050-ARoadmapfortheGlobalEnergySector_CORR.pdf.

Intergovernmental Panel on Climate Change. 2021. "Climate Change 2021: The Physical Science Basis." Switzerland: Intergovernmental Panel on Climate Change. https://www.ipcc.ch/report/ar6/wg1/downloads/report/ IPCC_AR6_WGI_SPM_final.pdf.

Intergovernmental Panel on Climate Change. 2022. "Climate Change 2022: Impacts, Adaptation and Vulnerability." Cambridge, UK and New York, NY, USA: Cambridge University Press. https://www.ipcc.ch/report/sixth-assess ment-report-working-group-ii/.

Irfan, Umair. 2019. "Fossil Fuels Are Underpriced by a Whopping $5.2 Trillion." *Vox*, May 17. https://www.vox.com/2019/5/17/18624740/fossil-fuel-sub sidies-climate-imf.

Kane, Michael Kobina and Stephnie Gill. 2022. "Green Hydrogen: A Key Investment for Green Transition." *World Bank Blogs*, June 23. https://blogs.wor ldbank.org/ppps/green-hydrogen-key-investment-energy-transition

Krugman, Paul. 2021. "Opinion | Who Created the Renewable-Energy Miracle? - The New York Times." *The New York Times*, August 17. https://www.nyt imes.com/2021/08/17/opinion/us-obama-renewable-energy.html.

Kweku Attafuah-Wadee, Lalen Lleander, Henrique Pacini. 2022. "Circular Economy Can Help Stem Biodiversity Loss". *United Nations Conference on Trade and Development*, April 19. https://unctad.org/news/blog-circular-economy-can-help-stem-biodiversity-loss

Langley, William. 2022. "Australia Passes Landmark Legislation to Cut Carbon Emissions." *The Financial Times*, September 8. https://www.ft.com/content/9201f832-2082-41fd-a58c-e8b426c32980.

Lindsey, Rebecca. 2020. "Climate Change: Atmospheric Carbon Dioxide." *Climate.Gov*. https://www.climate.gov/news-features/understanding-climate/climate-change-atmospheric-carbon-dioxide.

Ma, Mingming, Shun Wang, and Fengyu Wu. 2021. "COVID-19 Prevalence and Well-Being: Lessons from East Asia." 2021. https://worldhappiness.report/ed/2021/COVID-19-prevalence-and-well-being-lessons-from-east-asia/.

Mazzucato, Mariana. 2018. *The Value of Everything: Making and Taking in the Global Economy*, New York: PublicAffairs. ISBN 978-161039675-2.

Morton, Adam. 2020. "Inside the Liberal State Stepping into a Low-Emissions Future." *The Guardian*, March 6, sec. Australia news. https://www.theguardian.com/australia-news/2020/mar/07/inside-the-liberal-state-stepping-into-a-low-emissions-future.

Munasinghe. 2019. *Sustainability in the Twenty-First Century: Applying Sustainomics to Implement the Sustainable Development Goals*, 2nd ed. Cambridge University Press.

Mundy, Simon. 2022. *Race for Tomorrow: Survival, Innovation and Profit on the Front Lines of the Climate Crisis*. London: William Collins.

Parry, Ian W.H., Simon Black, and James Roaf. 2021. "Proposal for an International Carbon Price Floor among Large Emitters." *IMF Working Papers*, Staff Climate Note No. 2021/001, June 18.

Peter G. Peterson Foundation. 2022. "Budget Basics: National Defense." June 1. https://www.pgpf.org/budget-basics/budget-explainer-national-defense.

Picciotto, Robert, and Vinod Thomas. 2021. "Opinion: The Real Problem in the World Bank's doing Business Indicator." *Devex*, October 19. https://www.devex.com/news/sponsored/opinion-the-real-problem-in-the-world-bank-s-doing-business-indicator-101848.

Quah, Danny, and Vinod Thomas. 2022. "Building Resilience and Preventative Responses to Withstand Global Crises." *East Asia Forum*. https://www.eastasiaforum.org/2022/04/26/building-resilience-and-preventative-responses-to-withstand-global-crises/.

Raworth, Kate. 2017. The Doughnut of Social and Planetary Boundaries. https://www.kateraworth.com/doughnut.

REN21 Secretariat. 2022. "Renewables Global Status Report." France: United Nations Environment Programme. https://www.ren21.net/wp-content/uploads/2019/05/GSR2022_Full_Report.pdf.

Sachs, Jeffrey D. 2015. *The Age of Sustainable Development*. Columbia University Press. ISBN: 9780231173148. March.

Smil, Vaclav. 2022. *How the World Really Works: The Science behind How We Got Here and Where We're Going*. Viking.

Stern, Nicholas, Joseph E. Stiglitz, and Charlotte Taylor. 2021. "The Economics of Immense Risk, Urgent Action and Radical Change: Towards New Approaches to the Economics of Climate Change." *National Bureau of Economic Research*, NBER Working Paper Series, 28472. https://www.nber.org/system/files/working_papers/w28472/w28472.pdf.

Stiglitz, Joseph E., and Martine Durand, 2019. "Going Beyond GDP: Measuring What Counts for Economic and Social Performance". The Future of Economic Statistics. New York, March 1. https://unstats.un.org/unsd/statcom/50th-session/side-events/documents/20190301-1M-HLEG_Report_Friday_Seminar.pdf.

The White House. 2022. "National Strategy to Develop Statistics for Environmental Economics Decisions." Public Comment Draft – Federal Register Document ID 2022-17993 Regulations.gov Docket Number OMB-2022-0009. August 18.

The World Bank. 2021. The Changing Wealth of Nations: Managing Assets for the Future, October 27. https://openknowledge.worldbank.org/handle/10986/36400.

Thomas, Vinod. 1980. Welfare Cost of Pollution Control. *Journal of Environmental Economics and Management* 7 (2): 90–102. https://doi.org/10.1016/0095-0696(80)90011-X.

Thomas, Vinod, Mansoor Dailimi, Ashok Dhareshwar, Daniel Kaufmann, Kishor Nalin, Ramon Lopez, and Yan Wang. 2000. The Quality of Growth. *World Bank Publications*. https://doi.org/10.1596/0-1952-1593-1.

Thomas, Vinod. 2021. "To Support Climate Action, Growth Measures Should Count Planetary Damages." *Brookings* (blog). January 25. https://www.brookings.edu/blog/future-development/2021/01/25/to-support-climate-action-growth-measures-should-count-planetary-damages/.

Thomas, Vinod, and Namrata Chindarkar. 2019. *Economic Evaluation of Sustainable Development*. Singapore: Springer Singapore. https://doi.org/10.1007/978-981-13-6389-4.

United Nations. 2014. System of Environmental-Economic Accounting 2012—Central Framework. https://unstats.un.org/unsd/envaccounting/seearev/seea_cf_final_en.pdf.

United Nations Development Programme. 2022. "Planetary Pressures–Adjusted Human Development Index." *Human Development Reports*. United Nations. https://hdr.undp.org/planetary-pressures-adjusted-human-development-index.

United Nations Environmental Programme. 2018. Inclusive Wealth Report, September 21. https://www.unep.org/resources/inclusive-wealth-report-2018.

United Nations Office for Disaster Risk Reduction. 2022. Global Assessment Report on Disaster Risk Reduction. 2022. https://www.undrr.org/public ation/global-assessment-report-disaster-risk-reduction-2022.

United Nations Environment Programme. 2020. "Record Global Carbon Dioxide Concentrations despite COVID-19 Crisis." *UNEP*, 2020. http://www.unep.org/news-and-stories/story/record-global-carbon-dioxide-concen trations-despite-COVID-19-crisis.

Walljasper, Jay. 2011. "Elinor Ostrom's 8 Principles for Managing a Commmons | On the Commons." 2011. https://www.onthecommons.org/magazine/eli nor-ostroms-8-principles-managing-commmons.

Way, R, M.C. Ives, P. Mealy, and J.D.Farmer. 2022. "Empirically grounded technology forecasts and the energy transition." *Joule* 6 (9). September 13. https://doi.org/10.1016/j.joule.2022.08.009.

World Bank. 1992. "World Development Report 1992 : Development and the Environment." New York: Oxford University Press. https://openknowledge.worldbank.org/handle/10986/5975.

World Health Organization. 2018. "WHO: Health Benefits Far Outweigh Costs of Meeting Paris Goals | UNFCCC." *United Nations Climate Change*, 2018. https://unfccc.int/news/who-health-benefits-far-outweigh-costs-of-meeting-paris-goals.

Ziady, Hanna. 2020. "The Global Economic Bailout Is Running at $19.5 Trillion. It Will Go Higher." *CNN*, 2020. https://www.cnn.com/2020/11/17/economy/global-economy-coronavirus-bailout-imf-annual-report/index.html.

INDEX

A
Africa, 15, 66, 74–76, 110, 111
agriculture, 81
air pollution, 99, 106, 111, 112, 127
Amazon, 75, 98, 113, 129
anthropogenic, 115
Argentina, 25
Asia, 46, 74, 76, 77, 86, 111, 112, 161
Asian Development Bank (ADB), 110, 152, 158
Association of Southeast Asian Nations (ASEAN), 61, 80
atmosphere, 14, 18, 19, 21, 28, 73, 96, 98, 109, 133, 161
Australia, 44, 75, 108, 111, 112, 147, 177

B
Bangladesh, 145
behaviour, 183
Belt and Road Initiative (BRI), 86
biodiversity, 37, 98, 185
bottlenecks, 136

Brazil, 15, 25, 86, 110, 113, 188

C
Cambodia, 63
Canada, 25, 106, 110
carbon capture, 5
carbon neutrality, 27–29, 101, 131
carbon pricing, 3, 133, 155, 157, 178
carbon tax, 132, 138, 156, 157, 188
carbon taxation, 146, 156
Carmichael Coal Mine, 111
China, 5, 15, 85, 109–112, 132, 156, 188
climate finance, 5, 6, 27, 152, 154, 158, 163, 177
climate proofing, 4, 164
coal, 21, 57, 76, 85, 110–112, 116, 129, 136, 158, 159, 161, 180, 181
collective action, 55, 100, 101
Conference of Parties (COP), x, 101, 102
coping, 3, 24, 56, 61, 175, 182

© The Editor(s) (if applicable) and The Author(s), under exclusive license to Springer Nature Singapore Pte Ltd. 2023
V. Thomas, *Risk and Resilience in the Era of Climate Change*,
https://doi.org/10.1007/978-981-19-8621-5

COVID-19, 1, 2, 4, 5, 12, 17, 18, 22, 25, 26, 29, 38, 40, 54, 62, 66–68, 78, 80, 83, 98, 99, 102, 104, 106, 108, 112, 136, 150, 159, 160, 163, 175–181, 183, 184, 187, 189
cyber security, 5, 37, 38
cyclones, 13, 45, 48, 82, 145, 150, 162

D
Dallas, 77
debt, 18, 164
decarbonisation, 3, 75, 135, 144, 145, 157, 158, 182, 186, 187
deforestation, 3, 21, 38, 47, 49, 75, 98, 129, 186
Delhi/New Delhi, 106, 107, 130
developed countries, 78, 114, 163
developing countries, 5, 15, 18, 25, 114, 159, 160, 163, 177, 183
disaster management, 3, 16, 40, 48, 61, 84, 86, 87, 115, 144, 146, 149, 159
disaster risk management, 40, 47, 144, 161
disaster risk reduction, 40, 144, 149, 153, 156, 158, 162, 165
discounting, 131
droughts, 14, 21, 43, 47, 74, 81, 97, 98, 100, 108, 150, 151

E
early warning system, 53, 58, 146, 147, 165
earthquake, 13, 46, 49, 55, 57, 58, 83, 84, 147, 148, 162
East Asia, 17, 20, 76, 184
econometrics, 133
Economic Co-operation and Development (OECD), 54

ecosystem, 39
education, 3, 36, 37, 45, 46, 64, 78, 84, 87, 114, 129, 149, 154
effluent, 22, 28, 75, 100, 106, 111, 136, 156, 157, 179, 181, 183
electricity, 47, 48, 107, 136, 151, 158, 159, 161, 179
energy transition, 179
Europe, 15, 74, 76, 77, 125, 181
European Commission, 177, 180
European Union (EU), 13, 80, 109, 157, 177
evacuation, 45
evaluation, 188
exposure, 3, 40, 44–48, 53, 61, 62, 109, 112, 134, 135, 154, 161, 175, 189
externalities, 86, 126, 127, 129, 132, 133, 138, 157, 186, 188, 189

F
fires, 43, 44, 47, 100, 114, 115
floods, 11, 13–16, 20–22, 26, 43, 46, 47, 58, 59, 63, 74, 75, 77, 82, 97, 98, 100, 108, 111, 114, 115, 125, 135, 145, 148–150, 162, 184
food insecurity, 43
forest fires, 11, 14, 22, 43, 77, 98, 184

G
geopolitical conflicts, 17
Glasgow Climate Pact, 27
global governance, 103, 113
governance, 148
green financing, 161
green growth, 156
green hydrogen, 5

H
health system, 80
heatwaves, 11, 13–16, 47, 77, 97, 108
Hong Kong, 63, 80
hurricanes, 14, 16, 21, 74, 97
hydro-meteorological, 13, 135

I
impact evaluation, 150, 151, 188
India, 47, 82, 85, 86, 101, 106, 109–112, 150, 156, 161, 188
Indonesia, 58, 61, 110, 134
infrastructure, 146
intensity, 3, 12, 15, 20, 21, 44, 46–48, 56, 61, 79, 87, 189
Intergovernmental Panel on Climate Change (IPCC), 13, 27, 96, 101, 114, 131, 159, 175
International Monetary Fund (IMF), 65, 110, 111, 152, 158, 159, 178, 180, 181
intractability, 5, 6, 24, 97, 99, 101, 103, 115
Iran, 25, 110
irreversibility, 125, 131, 174
irrigation, 150

J
Japan, 101, 109, 110, 112, 146–148

K
Kerala, 67, 82, 85

L
landslides, 13, 47, 49, 148
Laos, 16, 63
Latin America, 74
Latin American, 74

logging, 75, 87, 98

M
mangroves, 126, 127
Mekong Delta, 63, 64
Mexico, 25, 61, 110
Middle East, 74, 76
migration, 36
mindset, 1, 2, 7, 55, 137, 148, 183
Montreal Protocol on Substances that Deplete the Ozone Layer, 104
Mozambique, 150

N
National Aeronautics and Space Administration (NASA), 13, 96
net zero, 179, 181
New Zealand, 54, 67, 78
Nigeria, 25

O
Odisha, 82
oil and gas, 15, 76, 109, 112, 174, 179–181
overshoot, 22, 174
Ozone, 105
ozone layer, 22, 99, 104, 134

P
Pakistan, 12, 15, 21, 26, 58, 74, 75, 125
Paris agreement, 27–29, 102, 131, 160, 161, 174, 187
Peru, 25, 61
Philippines, 14, 21, 82, 83, 111, 147, 149
Poland, 25
political economy, 5
political will, 3, 18, 159, 178, 187
power plant, 106, 110, 127, 184

preparedness, 3, 17, 29, 35, 41, 44, 45, 53, 55, 56, 62, 63, 78, 82, 87, 147–149, 175
prevention, 4, 35, 42, 44, 47, 56, 58, 61, 63, 78, 84, 148, 151, 182
private sector, 37, 64, 148, 163, 164, 181, 182, 189
public health, 67

Q
quality of growth, 185

R
renewable energy, 180
renewables, 156, 158, 159, 161, 179, 180
Russia, 1, 11, 17, 18, 65, 101, 109, 110, 156, 174, 179, 180, 188
Russia–Ukraine War, 65, 76

S
Saudi Arabia, 110, 188
Singapore, 16, 54, 63, 67, 78, 80, 112, 134, 149, 156, 157
social capital, 55
South Africa, 26
South Asia, 20
South China Sea, 17
Southeast Asia, 46, 101, 111, 126
Southeast Asia Health Pandemic Response and Preparedness, 80
South Korea, 25, 67, 78, 110, 149, 155
Spain, 47
spillover, 3, 38, 126, 127, 132, 137, 163, 165, 182
Sri Lanka, 150
storms, 15, 16, 21, 26, 43–47, 54, 63, 82, 83, 135, 148
super wicked problem, 5, 6, 103, 104

supply chain, 18, 39, 150
sustainable development, 7, 26, 98, 137, 149, 173, 175, 176, 185, 186, 188, 189
Sustainable Development Goals (SDGs), 86
Switzerland, 25
systemic risk, 39

T
Thailand, 63, 111, 126, 127, 150
the US, 15, 16, 25, 28, 59, 74, 76, 78, 101, 106, 108–110, 114, 136, 156, 157, 162, 178, 188
tipping points, 74, 125, 131, 174
transportation, 6, 106, 107
tsunami, 83, 84, 147, 148
Typhoon Haiyan, 14, 82, 147
typhoons, 148

U
Ukraine, 1, 11, 17, 18, 65, 174, 179
UN COP summits, 101
United Nations Framework Convention on Climate Change (UNFCCC), 21, 27
United Nations Office for Disaster Risk Reduction (UNDRR), 49, 175
urbanisation, 45, 153

V
Vietnam, 16, 63, 64, 111, 144–146
vulnerability, 3, 38, 40, 44–48, 53, 59, 61–63, 68, 83, 134, 135, 145, 151, 175, 189

W
waterlogging, 22

wicked problems, 5, 84, 88, 104, 116, 189
wildfires, 13, 15, 16, 21, 97, 114, 162
World Bank, 24, 64, 66, 81, 130, 137, 145, 150–152, 158, 159, 163, 185

Z
zoning, 45

CPSIA information can be obtained
at www.ICGtesting.com
Printed in the USA
BVHW051701170323
660678BV00002B/52